Critical Perspectives
on Nonacademic Science
and Engineering

Research in Technology Studies

Critical Perspectives on Nonacademic Science and Engineering

Research in Technology Studies,
Volume 4

EDITED BY

Paul T. Durbin

Bethlehem: Lehigh University Press
London and Toronto: Associated University Presses

Associated University Presses
440 Forsgate Drive
Cranbury, NJ 08512

Associated University Presses
25 Sicilian Avenue
London WC1A 2QH, England

Associated University Presses
P.O. Box 39, Clarkson Pstl. Stn.
Mississauga, Ontario
Canada L5J 3X9

The paper used in this publication meets the requirements
of the American National Standard for Permanence of Paper
for Printed Library Materials Z39.48-1984.

Library of Congress Cataloging-in-Publication Data

Critical perspectives on nonacademic science and engineering / edited
by Paul T. Durbin.
    p.   cm. — (Research in technology studies ; v. 4)
  Includes bibliographical references.
  ISBN 0-934223-15-7 (alk. paper)
  1. Technology—Philosophy.   I. Durbin, Paul T.   II. Series.
T14.C77   1991
601—dc20   89-64066
                                                                  CIP

PRINTED IN THE UNITED STATES OF AMERICA

Contents

6 *Contents*

Foreword

STEVEN L. GOLDMAN
STEPHEN H. CUTCLIFFE

This is the fourth volume in the series *Research in Technology Studies*, published by Lehigh University Press. Each volume consists of invited essays on a theme set by a guest editor. Paul Durbin has long argued that an understanding of the "real-world" contexts of the practice of applied science and engineering is central to understanding technology, either as a body of knowledge or as a social process. As guest editor of this volume, Durbin has invited a group of students of technology to address distinctive intellectual, social, and political issues that they perceive posed by science and engineering as practiced in nonacademic settings. The resulting collection of essays ranges very widely indeed, from a technical analysis of one facet of engineering reasoning to the politics of design. Taken together, the essays begin to define the parameters of an as yet virtually nonexistent discipline, namely, philosophy of engineering. Their publication will, we hope, spur a continuing conversation that will make philosophy of engineering part of the ongoing study of technology.

Acknowledgments

I owe special thanks for this volume to the coeditors of the series, Stephen Cutcliffe and Steven Goldman. Ever since the idea was first proposed they have encouraged me in every possible way; Steve Goldman worked with me in minute detail on choosing the precise outline and contributing authors, and both of them have waited patiently as one deadline after another passed. I also owe special thanks to Mary Imperatore and Dorothy Milsom, secretaries in the Philosophy Department at the University of Delaware, for what must at times have seemed like endless retypings of manuscripts. And I owe the most special debt of all to the corporate engineers and scientists with whom I have worked over the past twenty years in Delaware. They have shown me it is possible for people in large corporations to reflect philosophically on their roles and to dedicate themselves to the social responsibilities of science and engineering in a world plagued by technosocial problems. Finally, I would like to dedicate this volume to students looking forward to a career in R&D (including two of my daughters, Mary K. and Maggie, who seem headed in that direction); perhaps the more reflective approach suggested in this volume (and hopefully to be expanded in successors) can lead to more social responsibility in R&D settings, thus making them a better place to work.

PAUL T. DURBIN

Introduction

PAUL T. DURBIN

In the early 1960s, the Philosophy Department at the University of
Delaware hosted a well-respected series of conferences on the
philosophy of science—the Delaware Seminar—and was about to
launch an ambitious Ph.D. program. The graduate program never
materialized, and one reason given was that there was not a critical
mass of faculty. When I arrived at Delaware a few years later, I was
given another reason: the powerful local research-and-development
establishment had kept an eye on these events and decided that
academic analytical philosophy of science had nothing to offer to
them.

This volume addresses that issue squarely and focuses on the
everyday workplaces of the overwhelming majority of scientifically
and technically trained personnel (as reported, for instance, in
Science & Engineering Indicators, 1987[1]) and their analogues
throughout the developed world. That is, it is an attempt to shift
attention from pure science to the real world of the overwhelming
majority of scientists, engineers, and other technical specialists. In
other words, the volume focuses on all scientifically and technically
trained people *except* so-called pure scientists. This represents a
deliberate attempt to shift attention from the usual focus of academic interpreters of science—philosophers, historians (except for
some historians of technology), and sociologists of science. If there
is a theme for the volume, then, it is this focus on what often gets
neglected in studies of science, technology, and engineering.

THE NEED FOR A PHILOSOPHY OF R&D

When I first conceived this project, I had in mind a narrower
focus—the so-called R&D community. *Research and development*
as a descriptive phrase covering U.S. scientists, engineers, and

other technical personnel in industrial, governmental, and related laboratories seems to date from the decade after World War I— although a similar phenomenon, the application of engineering and science to the improvement of agriculture, had preceded it by a generation.[2] These institutions—industrial research, government research agencies and their satellite research institutions, and agricultural experiment stations and related institutions—do not complete the story. One should think also of military and space research, information and communications research, biomedical and pharmaceutical research centers, biotechnology ventures, applied research programs in academia, and economic and social science research centers to cover the full range of research-and-development institutions today.

This is not virgin territory, having been explored by at least one academic, Mario Bunge, who begins his most recent discussion of philosophy of technology with a clear definition:

> Technology may be conceived of as the scientific study of *the artificial* or, equivalently, as R&D (research and development). If preferred, technology may be regarded as the field of knowledge concerned with designing artifacts and planning their realization, operation, adjustment, maintenance and monitoring in the light of scientific knowledge.[3]

Bunge elaborates his definition of technology as a systematic set of eleven items that *necessarily* characterize any component in what he calls the "family of technologies." To this set, Bunge adds two assumptions: particular technologies so characterized are always necessarily related to other technologies, and they are always necessarily changing. For Bunge, technology is a *system* in the strict sense; he even thinks it ought to be understood in terms of *general systems theory*—if that theory is ever elaborated adequately.

My idea for the project was to make a new beginning, but even within the limits of traditional philosophy, there would be obvious points of disagreement with Bunge. In his discussion of the philosophical background of R&D work, Bunge makes statements that are calculated to provoke discussion from all sorts of philosophers—pragmatists, instrumentalists, positivists, realists, Marxists, idealists, and so on. In my view, if serious scholars come forward to debate a view such as that of Bunge, that is evidence enough that the view merits at least some attention. And this suggests that there are intellectual issues to be raised here. As a *praxis*-oriented philosopher, I would hope that the intellectual problems also represent real-world social problems. In this case, I would say there is an opening for a new field: philosophy of R&D. Others might see this as a subset of another relatively new field, philosophy of technology. I would not quibble with that claim, but I would add that, for me, the former is potentially the most interesting part of philosophy of technology.

This hardly demonstrates that among neglected aspects of academic treatments of science and engineering there is a need for a philosophy of R&D. But there are other instances of issues relevant to R&D that are already being debated—sometimes with only the dimmest awareness that such issues are relevant to the R&D community—and that could, if brought together, form the core of a philosophy of R&D.

Bunge is even more likely to be challenged on the value dimension of his system set—where he naively, critics would say, assumes that the values of technicians, including social engineers and "psychotechnologists" as well as high-technology medical practitioners, are superior to those of nonexperts. Bunge is an outspoken advocate of the "technology-is-neutral" view so often held by people in R&D:

> The primary responsibility for the evils of our time rests [not with science and technology in the abstract, but] with the ultimate political and business decision makers. Let us then blame them for the arms race and the unemployment it causes, for the acid rain and the destruction of the rain forests, for junk food and junk culture. The applied scientists and technologists involved in such processes are only accessories—alas sometimes unduly enthusiastic ones.[4]

The list of critics and opponents of the technology-is-neutral claim grows longer with every passing year. I will cite here only one critique, aimed clearly at Bunge. Joseph Margolis, attempting to establish a middle ground between the antitechnology pessimism of a Jacques Ellul and the naive optimism of Bunge, makes this strong claim:

> Bunge is utterly out of touch in supposing that the connection between technological power and its morally defensible use is straightforwardly one of means to ends. . . . Once . . . science is historicized, and science and technology praxicalized, there is no longer room for the elementary confidence Bunge exudes.[5]

I do not here want to claim that Margolis's praxicalized epistemology[6] is superior to Bunge's "epistemological realism with a touch of pragmatism." Indeed, I would take a different tack and, with George Herbert Mead, eschew the "riffraff" of epistemology in favor of a progressive social philosophy of the R&D community.[7] My point is, rather, that claims about the alleged objectivity or value-neutrality of R&D represent a battleground waiting to be fought over. It cannot be assumed—as Bunge seems to think—that even pure science is automatically progressive and needs no social or cultural legitimation.

These first two issues may, to defenders of the neutrality of R&D work, seem academic. Much more obviously controversial aspects of the R&D community are to be found in ethical and social problems. As one example, consider biotechnology. What I mean here is the whole range of products and processes—from technologically produced organs and prosthetic devices, through technologies such as kidney dialysis, to recombinant-DNA techniques and their applications, and ultimately to possibilities such as cloning humans—that have become possible by applying recent biological discoveries, especially in genetics, by way of R&D. I mean thus to include remote possibilities along with already-developed techniques and products.

I recognize at least five contributions to current philosophical disputes[8] on this topic:

1. At one extreme is the claim, voiced by some scientists, that the only control that can be ethically defended in this area is that by peers—scientific self-regulation. The assumption here is that the pursuit of truth is an inherent good with a high moral value, and the disinterested practice of science represents the pursuit of truth to an eminent, perhaps the preeminent, degree. And, by definition, government censorship or any other form of external control means interference with the disinterested pursuit of truth.

2. Biotechnological entrepreneurship, like any other form of free enterprise, should be subject only to the controls of a free market. The assumptions underlying this claim are complex: when people pursue their private ends, a common or social goal is achieved (the "hidden hand" in one version or another); the regulation of trade by government tends at least to lead to economically inefficient monopolies (history is alleged to be our evidence); and, in moral terms, private property, including its use for profit, is an inherent good with high value. A proviso is often added at this point: as long as use of the property does not unjustly harm others.

Probably a more widespread version of this claim is that biomedical science should be subject only to scientific self-regulation, but any profit-oriented uses of such research ought to be subject to the normal regulatory processes to which all commercial enterprises are subject in the contemporary welfare state. Here it is assumed that the free-enterprise system had led to many abuses before government regulation was applied, around the turn of the twentieth century; and, right down to the present, particular abuses have needed to be dealt with by government regulation. The moral principle most often associated with these claims is that the public has a right to governmental protection from entrepreneurs who, knowingly or unknowingly, abuse the privileges of the free-enterprise system. Most people who would take this view make the further assumption that, in itself, capitalism is a morally defensible economic system.

3. The Science for the People opponents in the recombinant-DNA debate of the 1970s seem to have been making a different claim, namely, that all biomedical science ought to be subject to democratic control—and especially such inherently risky ventures as recombinant-DNA research—and that science-for-profit is particularly suspect. Whether this is an accurate portrayal of the views of that particular group (or others like it), one thing is clear: there are Marxists who criticize capitalism as immoral and who are particularly suspicious of biotechnology in the hands of capitalists.

4. A fourth claim about bioengineering and biotechnology made on moral grounds is the suspicion of such research voiced by many religious thinkers (and echoed by other technology critics who, while not particularly religious, share the same mode of thinking): biotechnology is inherently dangerous and particularly so when it involves tinkering with nature—especially human nature. The assumption here is that there is something either God-given or quasi-divine about nature, and that humans tinker with it at their peril. This is often supported by claims about disasters associated with technological "advances"—such as the so-called green revolution or DDT. The most extreme version of this view is anti-science if not anti-intellectual—the belief that there are areas of nature that are off-limits to science, and even more so to technology.

5. This brings me to a final claim about biotechnology, one that I can support. I believe that biomedical engineering and biotechnological innovations of all sorts ought to be subject to democratic control; that is, I do not believe that regulation is needed only when potential risks are high (let alone only when the risks are out of hand). I do believe, however, that there ought to be levels of regulation. It seems to me, first, that with respect to research on genetic diseases (including therapies using DNA techniques), the

current system of oversight by ethics committees and similar bodies is adequate. Second, for industrial research—including agricultural, pharmaceutical, and biomechanical equipment manufacturing—I see nothing wrong with going beyond the current system of voluntary compliance with federal guidelines to a compulsory system. Finally, if it ever comes to proposed developments such as cloning humans, I hope we will have another full-scale public debate, just as we had in the recombinant-DNA controversy.

My rationale applies only to Western democracies—and perhaps only to the United States. I believe that, in our democratic system, *all* citizens should be equal in a moral sense; that, no matter what their social standing or level of intellectual sophistication, citizens know what is good for them and for those they love. This means, relative to the issue at hand, that in any conflict between groups— including a conflict between biotechnologists and people who would oppose their projects in particular cases—all citizens in a U.S.-style democracy have an equal right to protect the basic values that they feel are fundamental. In this matter of the fundamental right of all citizens *equally* to dissent against any and all policies, there is no place for expertocracy. There may be an important place for experts, but expertise should never preempt the basic rights of citizens.

The objection might be raised that this issue does not belong, in any significant way, to engineers and scientists in R&D. My response to this claim is based on my philosophical perspective— which is largely indebted to the thought of George Herbert Mead and John Dewey[9]—and others may not respond in the same way. It seems to me that if we accept the assumption of this objection, making a sharp distinction between scientists and engineers as professionals and as citizens, we undercut the meaning and purpose of science and engineering in modern society. To be a scientist or engineer, on my interpretation, is to be involved in community problem solving. Sometimes the people and their representatives in government may be satisfied with treating scientists and engineers as "hired hands," telling them to keep out of policy questions. But it is equally plausible that in a democratic society scientists and engineers should be held responsible for contributing to the solution of important social problems. Contemporary society is tolerant of biotechnologists' making a profit on recombinant-DNA research even though much of it has been federally funded, but as a society we also have every right to expect that this biomedical research will help solve some pressing health problems. More generally, if it is established that R&D contributes, however indirectly, to social problems of any sort, society can demand that members of the R&D community, *as such,* contribute to the solution of those problems.

To me, there are obvious problems associated with R&D that clamor for philosophical discussion. Some, as mentioned, are fairly traditional or academic—issues dealing with the epistemological assumptions that underlie the know-how of scientists and engineers working in R&D settings, or with the widespread claim by such scientists and engineers that their work is value-neutral. More urgent are social problems that should worry members of the R&D community as much as they do thinking people in all walks of life today—including, for instance, increasing economic inequities between high-tech haves and technological have-nots.[10] What I would argue is that this set of issues—epistemological, ethical, social, even economic—ought to constitute the core of a new sub-discipline, philosophy of R&D.

ASPECTS OF SCIENCE AND ENGINEERING NEGLECTED IN ACADEMIC SCHOLARSHIP

The authors I chose for the volume had ideas of their own—something easily understandable since almost every single one was already working on a larger project, and his or her contribution is derived therefrom. In almost every instance, what the authors have chosen to focus on is some aspect of science or engineering neglected by their colleagues in academia.

Deficiencies in Engineers' Self-Awareness

Billy Vaughn Koen begins the volume with a lament over the failure of engineers to articulate a philosophy of engineering that could compete with philosophy of science—so well known in large sectors of the educated public. Koen does not just lament the failure, however; he goes on to articulate a philosophy of engineering as a social problem-solving art with links to human problem solving in all areas of thought and action.

Edwin T. Layton, Jr., provides a historical demonstration of the slow process of emerging self-awareness, among engineering spokespersons, that engineering is not value-neutral but value-driven at every level and every stage. This new awareness on the part of the most enlightened members of the engineering professions, Layton shows, only came—after World War II—as a result of the reformulation of the design process in systems terms. In passing, Layton also addresses inadequacies in philosophy of technology—specifically in the work of Martin Heidegger. According to Layton, the historical record vindicates some aspects of Heidegger's views and challenges others. The most important lack in Heidegger, Layton says—and this affects many other philosophers of technology as well—is careful empirical documentation of the sort historians can supply.

Carl Mitcham begins with a set of data much the same as Layton's, that is, engineers' evolving self-descriptions. However, he goes on in a self-conscious attempt to formulate a strict definition of modern engineering in Aristotelian genus-and-species terms— namely, as *efficient human making* (where making falls under the more general category of art, and is distinguished from human *doing*—in ethics, politics, etc.). Mitcham notes how the cult of efficiency in modern engineering leads—as with the fine arts since the Renaissance—to a belief in the necessity of innovation, of ever-better (advocates would say) modes of production, instruments, and processes. In the end, Mitcham's essay is multiply critical: he charges Koen and other defenders of engineering with a failure to probe their profession's deepest meaning; but he also takes on critics of engineering. In particular, he points out how much of engineering ethics (not to mention technology assessment) offers no possibility of controlling technology because these academics ignore the inherently antitraditional character of engineering's commitment to innovation in the name of efficiency.

Deficiencies in Epistemologies of Technical Knowledge

The context of the two epistemological essays here is the recent turmoil in philosophy of science—the so-called Kuhnian revolution, culminating in recent radically historicist or radically relativist philosophies of science.

Ronald Laymon is among those philosophers of science who have resisted the new movements in philosophy of science, who have chosen instead to push ahead with the older tradition, applying it to ever-narrower scientific subspecialties. Laymon has gone even further, applying the logical tools of traditional philosophy of science methods to realms outside science entirely, namely, to thought processes in engineering. He is the only philosopher of science of note to do so.

Steven L. Goldman, is more sympathetic with contemporary

critics of the older tradition in philosophy of science. Nevertheless, Goldman targets deficiencies among a broader group of intellectuals. Goldman's focus is on trying to understand—even to unmask—what he sees as a two-part ideology of engineering. This ideology has characterized engineering as applied science, and *therefore* as value-neutral. This view has been accepted by all sorts of intellectuals in our culture, including philosophers of science but social critics of science and technology as well. They ignore, however, the point that engineering deserves attention on its own merits and not just as an extension of science. According to Goldman, engineering—and science in R&D settings—is "captive," and *must* be so. It is driven by largely arbitrary and idiosyncratic managerial decisions at every stage of the game. To people in R&D, this will come as no surprise, but that is not Goldman's point. What he means to highlight is the connection between this state of affairs and the neglect of engineering and "captive" science in our contemporary culture.

Value Dimensions of Technical Work

By now there is a fair amount of discussion, even among scientists and engineers themselves (not to mention their critics), of science, technology, and social responsibility. This is our third area of neglected opportunities.

Hans Lenk, in his very brief contribution (which is, it should be noted, a translation of a fraction of a much longer discussion in German), points out a major intellectual problem in claims about the social responsibilities of scientists and engineers. The concept of responsibility, even in famous declarations and codes of ethics of engineers, is often undefined and almost never differentiated in any hierarchical way. Lenk's essay, although it amounts to little more than a running commentary on three schematic outlines of levels and types of technical responsibility, does make two things clear: If claims about social responsibility are to have any impact on scientists and engineers, these claims will have to be carefully nuanced, dependent on types and levels of responsibility. Furthermore, moral responsibility (which also has several forms) is a special category— irreducible and in no way diminished by layers of hierarchies in large organizations.

Henryk Skolimowski attacks a shortcoming that he finds even greater. Although many critics lament the lack of moral leadership in our technological world, almost no one steps forward to provide it. The message of some who are labeled technological prophets (e.g., Jacques Ellul) is largely negative and pessimistic;

Skolimowski, with his Eco-Philosophy, dares to provide positive direction for particular technological ventures and for our age of technological development as a whole. Clearly his message is debatable, but he deserves credit for spelling out a direction for our technological age.

Science, Technology, and Politics

Since the Second World War, science, technology, and government has been a popular topic in certain circles; thus, one would not expect to find in it the same omissions as occur in the academic areas mentioned previously. But this turns out not to be the case.

Sheila Jasanoff does not attack the science and government literature directly; instead, she chooses as her starting point the authors in the "strong program" in sociology of science (and, lately and limitedly, of technology). In her view, this largely positive addition to the body of literature on science and society nonetheless ignores one of the main areas of intersection between society and scientific experts: the courts, either regular courts of law or administrative hearings and regulatory reviews. Clearly, this is a major concern on the part of scientists and engineers in industry, and those people have no doubt—whatever philosophers and social critics might think—that society keeps science under very tight control. Jasanoff's careful legal documentation shows that the courts do sometimes defer to scientific expertise, but this is by no means always the case; and there is a perennial source of potential conflict in the differing ways in which the courts and the technical community understand scientific evidence.

Richard Sclove's essay is, by contrast, quite radical. The focus of his critique is the almost-universal belief that the technological design process is necessarily elitist, a matter for experts. Sclove assumes that if technology is ever to be brought under effective democratic control—and he further assumes that this is not only good but necessary—then this elitism is going to have to be overcome. His essay, part of a larger project spelling out the philosophical foundations of a democratic technology, is chiefly devoted to providing examples of technological design ventures that have been genuinely democratic.

Deficiencies in Engineering Education

The title here is the title of one of the two papers in this section—that by Günter Ropohl. He is echoing literally thousands of la-ments, made over the last century (many of them emanating from prestigious engineering educators), that there is something radically wrong with engineering education. What is more or less standard about Ropohl's appeal for change is that engineering education ill prepares students for the broader interpersonal, social, and political dimensions of the work that they will be doing. What he adds to this dimension is a European background, plus a focus on the interdisciplinary aspect of the technosocial problem solving that engineering students must face in the modern world. What is unique about Ropohl's proposal for dealing with this issue is his focus on a *systems* approach. He is convinced that engineering students must see problems as systems and be prepared for this by systematic interdisciplinary teaching.

Taft Broome does not deal with engineering education directly; he is concerned, rather, with the preparation of the faculty—especially in science, technology, and society programs—who would be needed if Ropohl's suggestions were ever to be implemented in engineering schools. Science, technology, and society experts, Broome says (based on years of experience working with interdisciplinary committees), must be of a very special kind. They must be able to avoid the demands that professional career ladders in science and engineering (but also, he adds, in the humanities and social sciences) place on especially young researchers. The best STS faculty will consist of people trained in cross-disciplinary antispecialization, people who can learn to suspend critical judgments and look at issues in a cross-disciplinary perspective.

CONCLUSION

What advantages might there be in putting together these diverse critical reactions to oversights and omissions on the part of academic students of science and engineering? Would it have been a better volume if the authors had stuck to my proposed narrower focus on the R&D community? I believe, in the end, that the volume does serve a number of useful purposes. It is good to be reminded that engineers have not been sufficiently self-critical—or at least that their awareness of the value dimensions of their work has only arrived recently. It is useful to point out to philosophers of science that they have missed an interesting object of analysis in engineering, and in Goldman's "captive" science. It is salutary to remind advocates of social responsibility on the part of scientists and engineers that their cause may have to be more nuanced and complicated; and it is necessary for someone among such critics to step forward with a new ethic to lead technological development in positive directions. The literature on science, technology, and policy can benefit from real-world examples, both of legal cases involving science and of genuinely democratic technological designs. Finally, the literature on deficiencies in engineering education can stand the addition of two new proposals—one for a systems approach and another for specially trained STS faculty.

In summation, then, our authors (myself included) believe that the world of R&D scientists and engineers has been neglected by academic students of science and technology. We hope that in pointing out all this oversight we will be nudging students of science, technology, and society forward to fill at least some of the voids.

NOTES

1. National Science Board, *Science & Engineering Indicators, 1987* (Washington, D.C.: U.S. Government Printing Office, 1987). This is the eighth in an ongoing biennial series reporting on the state of science, engineering, and technology in the United States.

2. See Alexandra Oleson and John Voss, eds., *The Organization of Knowledge in Modern America, 1860–1920* (Baltimore: Johns Hopkins University Press, 1979).

3. Mario Bunge, *Treatise on Basic Philosophy,* vol. 7, *Epistemology and Methodology 3: Philosophy of Science and Technology* (Dordrecht: Reidel, 1985), p. 231.

4. Mario Bunge, "Can Science and Technology Be Held Responsible for Our Current Social Ills?" in *Research in Philosophy & Technology* (Greenwich, Conn.: JAI Press, 1984), 7:20.

5. Joseph Margolis, "Three Conceptions of Technology: Satanic, Titanic, Human," in *Research in Philosophy and Technology* 7:155–56.

6. Joseph Margolis, *Pragmatism without Foundations: Reconciling Realism and Relativism* (New York: Blackwell, 1986). This is the first of three projected volumes of Margolis's magnum opus, *The Persistence of Reality.*

7. Peter Berger and Thomas Luckmann, in *The Social Construction of Reality* (Garden City, N.Y.: Doubleday, 1966), have led the way in this project, though their work is subtitled *A Treatise in the Sociology of Knowledge,* and they claim not to be doing philosophy. For a good reconstruction of G. H. Mead's approach, see Hans Joas, *G. H. Mead: A Contemporary Re-Examination of His Thought* (Cambridge: MIT Press, 1985).

8. Sheldon Krimsky, in *Genetic Alchemy* (Cambridge: MIT Press, 1982), summarizes the issues—though without limiting himself to five viewpoints.

9. See, among John Dewey's works, *Democracy and Education* (New York: Macmillan, 1916), and *Liberalism and Social Action* (New York: Putnam's, 1935); also, Ralph W. Sleeper, *The Necessity of Pragmatism: John Dewey's Conception*

of Philosophy (New Haven: Yale University Press, 1986), and Gary Bullert, *The Politics of John Dewey* (Buffalo, N.Y.: Prometheus, 1983). G. H. Mead's basic idea is spelled out in his "Scientific Method in the Moral Sciences," in *Selected Writings,* ed. A. Reck (Indianapolis, Ind.: Bobbs-Merrill, 1964), pp.171–211; see also Joas, *G. H. Mead.*

10. Albert Borgmann is one of the few non-Marxist philosophers of technology to wrestle with this pressing issue of technological society. See his "Technology and Democracy," in *Research in Philosophy and Technology,* 7:211–28; and his *Technology and the Character of Contemporary Life* (Chicago: University of Chicago Press, 1984).

Contributors

TAFT H. BROOME, JR., Sc.D. (civil engineering, George Washington), M.S. (ethics in science and technology studies, RPI), teaches civil engineering at Howard University and is the author of numerous articles on engineering ethics. He is also chair of the ethics committee, American Association of Engineering Societies, and a member or board member of the Committee on Scientific Freedom and Responsibility (AAAS), Student Pugwash, and the Committee on Science and Society of Sigma Xi (among others).

PAUL T. DURBIN, Ph.D. (Aquinas Institute), teaches ethics of various scientific and technical professions at the University of Delaware and Thomas Jefferson University Medical College. Author or editor of twenty books—including *A Guide to the Culture of Science, Technology, and Medicine* (1980, 1984) and *Dictionary of Concepts in Philosophy of Science* (1988), as well as the annual publications of the Society for Philosophy and Technology—he is a member of the board of SPT, the National Association for Science, Technology, and Society, and *Science, Technology, & Human Values* (among others).

STEVEN L. GOLDMAN, Ph.D. (Boston University), is Andrew W. Mellon Distinguished Professor in the Humanities at Lehigh University and the author of a great many articles on the social relations of science and technology. From 1977 until 1988, he was director of the Science, Technology, and Society Program at Lehigh University.

SHEILA JASANOFF is Professor and Director of the Program on Science, Technology, and Society at Cornell University. She has published several books, including one just completed, *The Fifth Branch: Science Advisers as Policy-Makers* (forthcoming) and is working on another book that deals with science and the courts.

BILLY VAUGHN KOEN, D.Sc. (nuclear engineering, MIT), is professor of mechanical engineering at the University of Texas at

Austin and author of *Definition of the Engineering Method* (1985), from which his contribution here is excerpted. He is a diplomate of the Institut National des Sciences et Techniques Nucléaires (France); a member of numerous honorary organizations such as Tau Beta Pi, Phi Beta Kappa, and Sigma Xi; and the recipient of several teaching awards, including the Chester F. Carlson Award, a Minnie S. Piper Professorship, and a Standard Oil of Indiana Award for Outstanding Teaching.

RONALD LAYMON teaches philosophy at the Ohio State University and has been a research fellow at the Center for Philosophy of Science at the University of Pittsburgh. He has published extensively in philosophy of science and his contribution here is part of a larger project that extends that work to philosophy of engineering. He has received several research awards from NSF and was project director for *Enigma,* a widely adopted, computer-assisted logic course.

EDWIN T. LAYTON, JR. teaches history at the University of Minnesota. His book *The Revolt of the Engineers: Social Responsibility and the American Engineering Profession* (1971), won the Dexter Prize of the Society for the History of Technology and has recently been reissued. He is the author of a great many articles on the history of technology and is past president of the Society for the History of Technology.

HANS LENK, Ph.D., did graduate work in philosophy at Freiburg University, Kiel University, and the Technical University of Berlin—where he began his teaching career. He is currently professor and chairman of the philosophy department at the University of Karlsruhe. He has authored or edited over fifty books, including *Zur Sozialphilosophie der Technik* (1982), *Technik und Ethik* (1987), and *Technikbewertung* (1988). He is an Olympic rower, has published several books on sports, and has received numerous awards. He has also been vice president of the European Academy of Sciences and held many other prestigious positions.

CARL MITCHAM, Ph.D. (Fordham), teaches at the Pennsylvania State University and is director of the Philosophy and Technology Studies Center at Polytechnic University in New York. His books include the prize-winning *Bibliography of the Philosophy of Technology* (1973), *Philosophy and Technology* (1972, 1983)—both edited with Robert Mackey—*Theology and Technology* (1984), and

¿Que es la filosofía de la tecnología? (1989). He was the first president of the Society for Philosophy and Technology.

GÜNTER ROPOHL, Ph.D. (engineering), is Professor of General Technology at the University of Frankfurt and was previously director of the Institute for General Studies at the University of Karlsruhe. He has written widely on systems engineering, technology assessment, and the philosophy of technology; his books include *Eine Systemtheorie der Technik* (1979) and *Technik und Ethik* (1987, coedited with Hans Lenk).

RICHARD E. SCLOVE, Ph.D. (MIT), is executive director of the Loka Institute, a non-profit research organization in Amherst, Massachusetts, specializing in democratic prospects for science, technology, and architecture. He also has degrees in environmental studies and nuclear engineering and has published several articles related to his various research interests.

HENRYK SKOLIMOWSKI, D.Phil. (Oxford), did his original graduate studies in Warsaw. He is director of the Eco-Philosophy Center in Michigan, where he also teaches at the University of Michigan; in addition, he regularly runs workshops on ecophilosophy and teaches yoga all over the world, including Greece and India. His best known book is *Eco-Philosophy: Designing New Tactics for Living* (1981), but he has also published a great many other works, including *Technology and Human Destiny* (1983) and *Eco-Theology: Toward a Religion for Our Times* (1985).

Critical Perspectives
on Nonacademic Science
and Engineering

Part 1
Historical Background

The Engineering Method

BILLY VAUGHN KOEN

INTRODUCTION

Although philosophers and others use the term *technology* freely, it is unlikely that we have the vaguest notion philosophically of what it is or what is befalling us as it soaks deeper into our lives. Were we asked, "What is the scientific method?" we would undoubtedly answer without difficulty. We might propose, "Science is theory corrected by experiment," or the other way around. With a bit more probing, we might explain the scientific method by developing Popper's theory of falsification or Kuhn's theory of paradigm shifts. But when asked, anyone, whether lay person, scientist, or science historian, would feel qualified to give a cogent response. Now, as we sit immersed in the products of the engineer's labor, we must ask: What is the engineering method?

The lack of a ready answer is not surprising. Unlike the extensive analysis of the scientific method, little significant research to date has sought the philosophical foundations of engineering. Library shelves groan under the weight of books by the most scholarly, most respected people of history analyzing the human activity called science. Among many, many others one finds the work of the Ionian philosophers (Thales, Anaximander, and Anaximenes), where many feel the germ of the scientific method was first planted in the sixth century B.C.; of Aristotle in the *Organon;* of Francis Bacon in the *Novum Organum;* of René Descartes in *Discours de la Methode;* of Karl Popper in *The Logic of Scientific Discovery;* and of Thomas Kuhn in the *Structure of Scientific Revolutions.*

Remembering that high school students do not take courses in engineering; that the study of technology is not required for a

This is a condensation of my book, *Definition of the Engineering Method* (Washington, D.C.: American Society for Engineering Education, 1985), aided considerably by editor Paul Durbin; the material here is used with the permission of the American Society for Engineering Education.

33

liberal arts degree; and that sociologists, psychologists, historians and religious proponents—not engineers—write most of the pro- and antitechnology literature, can we be sure, as the engineer speaks of optimization, factors of safety and feedback, that the lay person would understand an engineering spokesperson if one did exist? Not only is there little research into the theory of engineering, no recognized spokesperson and no general education requirements in the field, but engineers themselves are chronically averse to writing about their world. That people do not understand the engineering method and are a bit frightened by technology is thus not surprising.

THE ENGINEERING METHOD

Chess is a complicated game. Although in theory a complete game tree can be constructed by exhaustive enumeration of all the legal first moves for the white side, then all possible responses to each by black, then white again, and so forth until every possible game appears on the tree, in practice this procedure is impossible because of the enormous number of different moves and the limited resources of even the largest computer. Chess, therefore, defies analysis.

To further an understanding of chess, a different strategy is usually needed, one that is consistent with the available resources. This strategy consists of giving suggestions, hints, and rules of thumb for sound play. For example,

1. Open with a center pawn.
2. Move a piece only once in the opening.
3. Develop the pieces quickly.
4. Castle on the king's side as soon as possible.
5. Develop the queen late.

As we get better, we begin to hear,

6. Control the center.
7. Establish outposts for the knights.
8. Keep bishops on open diagonals.
9. Increase your mobility.

Although these hints do not guarantee a win, often offer conflicting advice, and depend on context and change in time, they are obviously better than trying to construct the game tree, and they do teach one to play better chess. What is the name of this strategy?

One of the topics studied in a course in artificial intelligence is an unusual way of programming a computer to solve problems. Instead of giving it a program with a fixed sequence of deterministic steps to follow—an algorithm, as it is called—the computer is given a list of random suggestions, hints, or rules of thumb to use in seeking the solution to a problem. The hints are called *heuristics,* the use of these heuristics, *heuristic programming.* Surprisingly, this vague, nonanalytic technique works. It has been used in computer codes that play championship checkers, identify hurricane cloud formations, and control nuclear reactors. Like the computer, both the method for solving its problem (learning to play chess) and that of engineers in solving their problems (building bridges, and so forth) depend on the same strategy for causing change. This common strategy is the use of heuristics. In the case of the engineer, it is given the name engineering design.

To analyze the important relationship between engineering design and the heuristic, four major objectives are set for this part of the discussion. They are to understand the technical term *heuristic;* to develop the engineer's strategy for change; to define a second technical term, *state of the art;* and finally, to state the principal rule for implementing the engineering method. The heuristic will be considered by definition, by examining its characteristics or signatures, and through its synonyms. The state of the art will be explained by definition and by looking at its evolution and transmission from one generation of engineers to the next. Five examples proving the usefulness of this important engineering concept will then be reviewed.

1. The Heuristic: A Definition

A heuristic is anything that provides a plausible aid or direction in the solution of a problem but is in the final analysis unjustified, incapable of justification, and fallible. It is used to guide, to discover, and to reveal. This statement is a first approximation, or as the engineer would say, a first cut, at defining the heuristic.

2. Signatures of the Heuristic

Although difficult to define, a heuristic has four signatures that make it easy to recognize:

- A heuristic does not guarantee a solution;
- It may contradict other heuristics;
- It reduces the search time in solving a problem; and
- Its acceptance depends on the immediate context instead of an absolute standard.

Let us compare the presumably known concept of a scientific law with the less-well-known concept of the heuristic with respect to these four signatures. In doing so we will come to appreciate the rationality of using irrational methods to solve problems.

This comparison may be easier if we use the simple mathematical concept of a set. The interior of figure 1 represents the set of all problems that can ultimately be solved. It will be given the name U. U is not limited to those problems solvable on the basis of present knowledge, but it includes all problems that would theoretically be solvable given perfect knowledge and an infinite amount of time. The points labeled, a, b, c, d, g, and h are elements of U and represent some of these problems. If you prefer, it is sufficient for present purposes to think of U as a simple list of questions about nature that humanity has answered or will someday be able to answer. On this list most people born into the Western tradition would include: Will the sun rise tomorrow? Does bread nourish? If I release this ball, will it fall? And, should the Aswan High Dam have been built? Outside this area is everything else—questions that humanity cannot answer, questions that humanity cannot even ask, and pseudoquestions. Many scientists believe that no points, such as e and f, exist in this outer region. This picture admittedly leaves unidentified and certainly unresolved many important issues.

In figure 1, the problems are encircled by closed curves labeled A through I. Similar to the dotted rectangle, the area inside each curve represents a set. Those that have been crosshatched, A, B, B', C, D, and I, are sets of problems that may be solved using a specific scientific or mathematical theory, principle, or law. Set A, with problem a as a representative element, might be the collection of all problems solvable using the law of conservation of mass-energy, and set B, those requiring the associative law of mathematics. If the area inside a curve is not crosshatched, such as E, F, G, and H, it represents the set of all problems that may be attacked using a specific heuristic. This figure helps illustrate the difference between a scientific law and a heuristic based on the four signatures given earlier.

First, heuristics do not guarantee a solution. To symbolize this

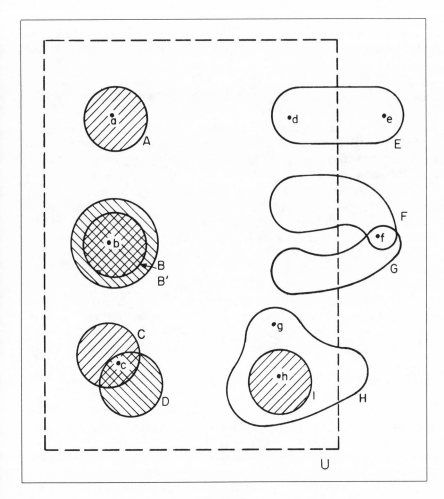

Figure 1

characteristic, the sets referring to scientific laws rest completely within set *U*, while those referring to heuristics include the area both inside and outside of *U*. When heuristic *E* is applied to problem *d*, a satisfactory solution results. This is not the case when the same heuristic is applied to problem *e*. To scientists, ambiguity about whether an answer to a question has been found is a fatal weakness. They seek procedures, strategies, and algorithms that give predictable results known to be true. Uncertainty about a solution's validity is a sure mark of the use of a heuristic.

Unlike scientific theories, two heuristics may contradict or give different answers to the same question and still be useful. This blatant disregard for the classical law of contradiction is the second sure signature of the heuristic. In figure 1 the overlap of the two scientific sets, C and D, indicates that a problem in the common area such as c would require two theories for its solution. The need for both the law of gravitation and the law of light propagation to predict an eclipse is a good illustration. Since combinations of two, three, and often more scientific and mathematical theories must work together to solve most problems, U is overlaid with a complex array of scientific sets.

This is not true in the case of the heuristic. Here, the overlap of F and G represents the conflicting answer given to problem f, found outside of U. Although at times two heuristics might be needed to arrive at an answer and hence to overlap within U, the most significant characteristic of a heuristic is its rugged individualism and tendency to clash with its neighbors. At least three different heuristic strategies are available that compute the number of Ping-Pong balls in a room, and each leads to a different, but completely acceptable, engineering answer. For a mathematician, contradiction is worse than ambiguity. Mathematicians might tolerate a heuristic strategy that indicates the direction to a solution if independent confirmation exists that such a solution represents the truth. A contradiction, however, is always unacceptable, for it implies a complete breakdown in the system. Logically, from any two propositions that contradict, any proposition at all may be proved to be true—certainly a bothersome situation in science and mathematics. Unlike scientific laws, heuristics have never taken kindly to the harness of conventional logic systems and may be recognized when they bridle.

Some problems are so serious and the appropriate analytical techniques to solve them either nonexistent or so time-consuming that a heuristic solution is preferable to none at all. Problem g in figure 1 is not a member of any crosshatched set, but it is a member of the heuristic set H. If g is lethal to the human species on a time scale shorter than that needed to develop a scientific theory to solve it, then the only rational course is to use the irrational heuristic method. Problem h represents a variant of this situation. It is a member of both H and I, but now let us assume that the time needed to implement the known, rigorous solution is longer than the lifetime of the problem. Again, first aid in the field is better than a patient dead on arrival at the hospital.

Unfortunately, most serious problems facing humankind are similar to g and h. Sufficient analytical theory or enough time to implement known theory does not exist to solve the problems of war, energy, hunger, and pollution. But in each case first aid in the form of heuristics is surely available—if only we knew how to use it.

Even though heuristics are nonanalytic, often false, and sometimes contradictory, they are properly used to solve problems so complex and poorly understood that conventional analytical techniques would be either inadequate or too time-consuming. This ability to solve unsolvable problems or to reduce the search time for a satisfactory solution is the third characteristic by which a heuristic may be recognized.

The final signature of a heuristic is that its acceptance or validity is based on the pragmatic standard—it works or is useful in a specific context—instead of on the scientific standard—it is true or consistent with an assumed, absolute reality. For a scientific law the context or standard of acceptance remains valid, but the law itself may change or become obsolete; for a heuristic the contexts or standards of acceptance may change or become obsolete, but the heuristic itself remains valid. Figure 1 helps make this distinction.

Science is based on conflict, criticism, or critical thought, on what has been called the Greek way of thinking. A new scientific theory, say B', replaces an old one, B, after a series of confrontations in which it is able to show that—as an approximation to reality—it is either broader in scope or simpler in form. If two scientific theories, B and B', predict different answers to a question posed by nature, at least one of them must be wrong. In every scientific conflict there must be a winner. The victor is declared the best representative of "the way things really are" and the vanquished discarded as an interesting but no longer valid scientific relic. Ironically, the loser is often demoted to the rank of a heuristic and still used in cases of expediency. Thus, Einstein's theory replaced Newton's as scientific dogma, and Newton's Law of Gravitation is now used, in the jargon of the engineer, when a *quick and dirty* answer is needed. The scientist assumes that the set U exists, that it does not change in time, that it is eternal. Only the set of currently accepted scientific laws changes in time.

On the other hand, the absolute value of a heuristic is not established by conflict but depends exclusively on its usefulness in a specific context. If this context changes, the heuristic may become uninteresting and disappear from view, awaiting, perhaps, an eventual change of fortune. Unlike a scientific theory, a heuristic never dies; it just fades from use. A different interpretation of figure 1 is therefore more appropriate in the case of a heuristic.

For the engineer the set U represents all problems he wants to answer at a given moment instead of all problems that are ultimately answerable. As a result, it is not a constant but varies in time. The engineer's set U ebbs, as the obsolescence of the buggy has left the heuristics for buggy whip design high and dry on the shelf in the blacksmith's workshop, and it flows as renewed interest in self-sufficiency has sent young people in search of the wisdom of the pioneers. One heuristic does not replace another by confrontation but by doing a better job in a given context. Both the engineer and Michelangelo "criticize by creation, not by finding fault."

The dependency on immediate context instead of absolute truth as a standard of validity is the final hallmark of a heuristic. It and the other three signatures are not the only important distinctions between the scientific law and the heuristic, but they are sufficient, I think, to indicate a clear difference between the two.

3. Synonyms for the Heuristic

Most engineers have never consciously thought of the formal concept of the heuristic, but all engineers recognize the need for a word to fit the four signatures. They frequently use the synonyms rule of thumb, intuition, technique, hint, rule of craft, engineering judgment, working basis, or, if in France, le pif (the nose), to describe this plausible, if fallible, basis of the engineer's strategy for solving problems. Each of these terms captures the feeling of doubt characteristic of the heuristic.

This completes consideration of the technical word heuristic needed for a definition of the engineering method; later, an extensive list of examples will be given. We have analyzed this important concept through analogy with the hints and suggestions given to learn chess, by definition, by looking at four signatures that distinguish it from a scientific law and by reviewing a list of its synonyms.

I hasten to add that neither the word heuristic nor its application to particularly intractable problems is original with me. Some historians attribute the earliest mention of the concept to Socrates (born about 469 B.C.), and others identify it with the mathematician, Pappus, around 300 A.D. Principal among its later adherents have been Descartes, Leibnitz, Bolzano, Mach, Hadamard, Wertheimer, James, and Koehler. In more recent times, George Polya (How to Solve It [Princeton: Princeton University Press, 1945, 1973]) has been responsible for its continued development. Without a doubt, the study of the heuristic is very old. But as old as it is, the use of heuristics to solve difficult problems is older still. Heuristic methods were used to guide, to discover, and to reveal a plausible direction for the construction of dams, bridges, and irrigation canals long before the birth of Socrates.

DEFINITION OF THE ENGINEERING METHOD

What is original in our discussion is the definition of the engineering method as the use of engineering heuristics to cause the best change in a poorly understood situation within the available resources. This definition is not meant to imply that the engineer just uses heuristics from time to time to aid in his work, as might be said of the mathematician. Instead, my thesis is that the *engineering strategy for causing desirable change in an unknown situation within the available resources* and the *use of heuristics* is an absolute identity. In other words, everything the engineer does in his role as engineer is under the control of a heuristic. Engineering has no hint of the absolute, the deterministic, the guaranteed, the true. Instead, it fairly reeks of the uncertain, the provisional, and the doubtful. The engineer instinctively recognizes this and calls his ad hoc method "doing the best you can with what you've got," "finding a seat-of-the-pants solution," or just "muddling through."

STATE OF THE ART

Instead of a single heuristic used in isolation, a group of heuristics is usually required to solve most engineering design problems. This introduces the second important technical term, *state of the art*. Anyone who has spent any time in the presence of an engineer likely has heard this term. The engineer might boast of personal possessions such as a state-of-the-art speaker system or a state-of-the-art computer design. Since this concept is fundamental to the art of engineering, let me turn, now, to the definition, evolution, and transmission of the state of the art, giving examples of its use.

1. A Definition

State of the art, as a noun or an adjective, always refers to a set of heuristics. Since many different sets of heuristics are possible, many different states of the art exist, and to avoid confusion each should carry a label to indicate which one is under discussion. Each set, like milk in the grocery store, should also be dated with a time stamp to indicate when it is safe for use. Too often, the neglect of the label and time stamp has caused mischief. With these two exceptions, no restrictions apply to a set of heuristics for it to qualify as state of the art.

In the simplest, but less familiar, sense, *state of the art* is used as a noun referring to the set of heuristics used by a specific engineer to solve a specific problem at a specific time. The implicit label indicates the engineer and the problem, and the time stamp shows when the design was made. For example, if an engineer wants to design a bookcase for an American student, he or she calls on the rules of thumb for the size and weight of the typical American textbook, on engineering experience for the choice of construction materials and their physical properties, and on standard anatomical assumptions about how high the average American student can reach, and so forth. The state of the art used by this engineer to solve this problem at this moment is the set of these heuristics. If the same engineer were asked to design a bookcase for a French student, he or she would use a different group of heuristics and hence a different state of the art. (Bookcase design is not the same in the United States and France.) Now consider two engineers who have been given the same problem of designing a bookcase for an American student. Each will produce similar, but different, designs. Since a product is necessarily consistent with the specific set of heuristics used to produce it, and since no two engineers have exactly the same education and past experience, each will have access to similar, but distinctly different, sets of heuristics and hence will create a different solution to the same problem. State-of-the-art as a noun refers to the actual set of heuristics used by each of these engineers.

In a complicated but more conventional sense, *state of the art* also refers to the set of heuristics judged to represent "best engineering practice." When a person says that his stereo has a state-of-the-art speaker system or that he has a state-of-the-art bookcase, he does not just mean that they are consistent with the heuristics used in their design. That much he takes for granted. Instead, he is expressing the stronger view that a representative panel of qualified experts would judge his speaker system or his bookcase to be consistent with the best set of heuristics available. Once again, state of the art refers to an identifiable set of heuristics.

Because the design of a bookcase requires only simple, unrelated heuristics, it misrepresents the complexity of the engineering state of the art needed to solve an actual problem. More typically, the state of the art is an interrelated network of heuristics that control, inhibit, and reinforce each other. For example, the physical property of a large organic molecule called the *enthalpy* may be determined either in the laboratory (one heuristic) or by estimating the number of carbon and hydrogen atoms the macromolecule contains and then applying a known formula (a rival heuristic). In practice, the engineer uses sometimes one method, sometimes the other. Obviously he has another heuristic—perhaps something like "go to the laboratory if you need 10 percent accuracy and have $5,000"—to guide his selection between the two. Bookcase design, and even the determination of the enthalpy of a large organic molecule, are such simple examples that they hardly suggest the complexity of a state of the art. It is to your imagination I must finally turn to visualize how much more complicated the state of the art used in the design of an airplane must be as the heuristics of heat transfer, economics, strength of materials, and so forth influence, control, and modify each other. Whether it is the set that was actually used in a specific design problem or the set that someone feels would be the best, state of the art always refers to a collection of heuristics—most often a very complicated collection of heuristics at that. Its imaginary label must indicate which engineer, which problem, and which set are under consideration.

The state of the art is a function of time. It changes as new heuristics become useful and are added to it and as old ones become obsolete and are deleted. The bookcase designed for a Benedictine monk today is different from the one designed for St. Benedict in 530 A.D. When discussing the design of a bookcase for students, I did not emphasize the time stamp that must be associated with every state of the art. Now is the time to correct this omission and consider the evolution of a set of heuristics in detail, beginning with a well-documented example.

2. Evolution

In overall outline, scholars feel that the evolution of the present-day cart took place in a series of stages. Since the wheel probably evolved from the roller, it is assumed that the earliest carts had wheels rigidly mounted to their axles, with wheels and axle rotating as a unit. This assembly has the disadvantage of forcing one wheel to skid in going around a corner. As an improvement, the second state was probably a cart with the axle permanently fixed to the body and with each wheel rotating independently. The disadvantage of this design was that neither the wheels nor the axle were capable of pivoting. Thus, the cart could not go around corners unless forced into a new position. The same difficulty bothers parents who push an old-fashioned baby stroller with fixed wheels.

After twenty or thirty centuries, the engineer learned how to correct this problem by allowing the front axle to pivot on a king bolt. Since the front and back wheels were large and the same size, this cart could not turn sharply without the front wheels scraping against its body. In the fourth and final stage, the front wheels were reduced in size and thus allowed to pass under the bed of the wagon as the front axle pivoted. The evolution of cart design continues in the present day, of course, the cart having become the automobile.

3. Transmission

Through the ages the state of the art has been preserved, modified, and transmitted from one individual to another in a variety of ways. The earliest method of transmission was surely a simple apprentice system, in which artisans carefully taught rules of thumb for firing clay and chipping flint to assistants who would someday replace them. With hieroglyphics, cave paintings and, later, books, the process became more efficient and no longer depended on a direct link between teacher and taught. Finally, in more recent times, trade schools and colleges began to specialize in teaching engineers. In spite of the importance of apprenticeship, books, and schools, in preserving, modifying, and transmitting accumulated engineering knowledge, they need not detain us longer because of their familiarity.

An additional method is less well known and worth note. If an engineer were asked to design a cart today, he or she would use an articulated front axle and wheels that were small enough to pass under the bed of the cart—not because of personal familiarity with the evolution of cart design, but because carts today are constructed this way. All traces of the state of the art that dictated axle and wheels turning as a unit have disappeared. The engineer does not need to know the history of cart design; the cart itself preserves the state of the art that was used in its construction. In other words, modern design does not recapitulate the history of ancient design.

Since the heuristics of tomorrow are embodied in the concrete objects of today, engineers are usually sensitive to the physical world around them and use this knowledge in their designs. I once asked an architectural engineer to estimate the size of the room in which we had been sitting during an earlier meeting. He quickly gave answer by remembering that the room had three concrete columns along the side wall and two along the front. Since he also knew the rule of thumb for standard column spacing that applied to a room such as ours, he could calculate its size quite accurately. This awareness of the present world translates directly into the heuristics used to create a new one. If a proposed room, airplane, reactor, or bridge deviates too far from what is expected, the engineer will question, recalculate, and challenge.

To review: the state of the art is a specific set of heuristics designated by a label and time stamp. It changes in time and is passed from engineer to engineer either directly, by the technical literature, or in a completed design. Typically, it includes heuristics that aid directly in design, those that guide the use of other heuristics, and, as we will see later, those that determine an engineer's attitude or behavior in solving problems. The state of the art is the context, tradition, or environment (in the broadest sense) in which a heuristic exists and based upon which a specific heuristic is selected for use. We might also characterize the state of the art of the engineer as his privileged point of view.

State of the art, no matter how it is written, is both a cumbersome and inelegant term. After only a few pages of definition, its use has become tiresome. Therefore, I propose to replace it with the acronym, *sota*. Coining a new word by using the first letters in an expression's constituent words is a familiar procedure to the engineer, who speaks freely of *radar* (radio detecting and ranging) and the *lem* (lunar excursion module). In large engineering projects such as the manned landing on the moon, acronyms are so frequently used that often a project description appears to the non-engineer to be written in a foreign language. At times an acronym even takes on a life of its own, and we forget the words used to create it. Few remember the original definitions of *laser, scuba, zip code,* and *snafu,* and I wish the same fate for *sota*. From now on, *sota,* used both as adjective and noun, is to be taken as a technical term meaning an identifiable set of heuristics.

4. Sample Uses of an Engineering Sota

Without due consideration, the concept of an engineering state of the art as a collection of heuristics appears contrived, and its acronym, gimmicky. The frequency with which the word *sota* will appear in the remainder of this discussion and its obvious advantage over the expanded form will answer the second criticism. Five specific examples showing the effectiveness of the sota as a tool for bringing understanding to important aspects of the engineering world will answer the first. Various sets of heuristics will now be used to

1. compare individual engineers,
2. establish a rule for judging the performance of an engineer,
3. compare the technological development in various nations,
4. analyze several pedagogical strategies of engineering education, and
5. define the relationship between the engineer and society.

This last example will also suggest the importance of technological literacy for the nonengineer and liberal literacy for the engineer. Although these examples are discussed at length, they should dispel any doubts about the legitimacy of a sota and suggest important areas for future research. We will then, at last, be in a position to consider the principal rule for implementing the engineering method.

COMPARISON OF ENGINEERS

The individual engineer, in his role as engineer, is defined by the set of heuristics he uses in his work. When his sota changes, so does his proficiency as an engineer. This will be represented in figure 2.

No two engineers are alike. The sotas of three engineers, *A, B,* and *N,* are represented in figure 2. They share heuristics inside the area where they overlap, but each engineer also encloses additional area to account for the unique background and experience of the individual it represents. In general, if *A, B,* and *N* are all civil engineers, the overlaps of their sotas is larger than if *A* is a civil engineer, *B* is a chemical engineer, and *N* is a mechanical engineer. Most civil engineers have read the same journals, attended the same conferences, and quite possibly used the same textbooks in school. Not surprisingly, they share many engineering heuristics.

If the sotas of all modern civil engineers, instead of only the three, were superimposed, the common overlap of all these sets, technically called the intersection, would contain the heuristics required to define a person as a modern civil engineer. Both society and the engineering profession have a vested interest in preserving

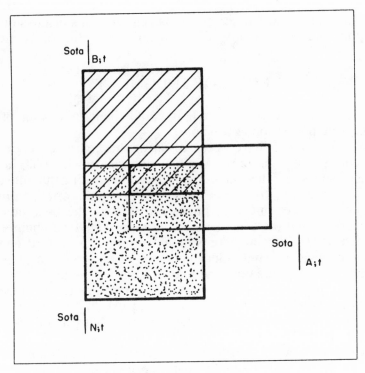

Figure 2

the integrity of this area. The label, *civil engineer,* must insure a standard approach and minimum level of competence in solving civil engineering problems. What is now needed is research to determine the minimum intersection necessary to certify an engineer as an expert in, say, heat transfer, hydraulics, or strength of materials. We will return to the intersection of all engineering sotas later in our discussion to give the heuristics it must contain for a person to be properly called an engineer. For now, remember that the overlap of the appropriate sotas measures the similarity and dissimilarity of engineers. It is an instructive example of the use of the concept sota.

No one, I think, would argue that engineers should not be held responsible for their work. The difficulty is knowing when they have done a satisfactory job. A discussion of the correct rule for judging the engineer will be a good second example of using an engineering sota and will emphasize, once again, the importance of attaching an imaginary label and a time stamp to each one. This discussion will also point out the difficulty of implementing this Rule of Judgment.

RULE OF JUDGMENT

If figure 2 were generalized, it could depict the sota of the engineering profession as a whole.

No engineer will have acess to all of the heuristics known to engineering, but in principle, some engineer somewhere has access to each heuristic. The black solid circles represent the subset of heuristics needed to solve a specific problem. The combined wisdom of the engineering profession defines the best possible engineering solution. This overall sota represents best engineering practice and is the most reasonable practical standard against which to judge the individual engineer. It is a relative standard instead of an absolute one, and like all sotas it changes in time.

To my knowledge, no engineers are clairvoyant. Handicapped in this way, it would seem unreasonable to expect them to make a decision at one moment based on information that will only become available later. They can only make a decision based on the set of heuristics that bears the time stamp certifying its validity at the time the design must be made. With these considerations in mind, we can formulate the fundamental Rule of Judgment in engineering: Evaluate an engineer or an engineering design against the sota that defines best practice at the time the design was made.

This rule is logical, defensible, and easy to state. Unfortunately, it is not universally applied because of ignorance, inattention, and a genuine difficulty in extracting the sota that represents good engineering practice from the set of all engineering heuristics in a specific case. Each of these three reasons for not observing the Rule of Judgment is worthy of special attention.

For the layperson, the failure of an engineering product usually means that some engineer has done a poor job of design. This criticism is based on ignorance of the correct basis for judging the engineer and is indefensible for two reasons. First, an engineering design always incorporates a finite probability of failure. The engineer uses a complex network of heuristics to create the new in the area of uncertainty at the margin of solvable problems. Hence, some failures are inevitable. Had ancient engineers remained huddled in the security of the certain, they would never have ventured forth to create the wheel or the bow. The engineer should not be criticized by looking only at a specific failure and ignoring the context or sota upon which decisions leading to that failure were based. Second, the world changes around the completed engineering product. A sota tractor on the date it is delivered is not necessarily appropriate for working steeper terraces and pulling heavier loads fifty years later; it may fail or overturn or both. Often the correct basis for judging the performance of an engineer is not used because the public, including juries charged with deciding product liability cases and journalists reporting major technological failures, does not know what that basis is.

Engineers are also misjudged through inattention. Since the sota is a function of time, special attention is needed to ensure that engineers are evaluated against the sota valid at the time they make their designs. Two examples will demonstrate how easy it is—even for engineers—to forget this requirement.

One of my French engineering colleagues, undoubtedly carried away with nationalistic zeal, surveyed the modern Charles de Gaulle Airport in Paris and explained that he had just had a miserable time getting through O'Hare Airport in Chicago. He then added that he could never understand why American engineers are not as good at designing airports as are the French. His mistake was wanting a clairvoyant engineer. Leaving aside the factor of scale— the American airport was at the time the busiest in the world; de Gaulle had been in operation only two weeks—two facts are beyond dispute: each airport was consistent with the sota at the time it was built and, given the intimate exchange of technical information at international technical meetings, the sota on which the French airport was based was surely a direct outgrowth of the earlier one used in Chicago. Even an engineer sometimes forgets the time stamp required on an engineering product.

The second example concerns the aphorism of the American frontier that a stream renews itself every ten miles. Essentially, this means that a stream is a buffered ecosystem capable of neutralizing the effects of an incursion within a short distance. Let us assume that an enterprising pioneer built a paper mill on a stream and into it discharged his waste. According to the above rule of thumb, no damage was done. Now let us add that over subsequent decades additional mills were constructed until the buffering capacity of the stream was exceeded and the ecosystem collapsed. Your job is to fix blame. If you argue that later engineers were wrong to use a heuristic that was no longer valid, I might agree. Presumably the original heuristic is applicable only to a virgin stream. If you argue that the original plant should be modified to make it consistent with later practice, a process the engineer calls retro- or backfitting, I might also agree, although I am curious as to whether you would require the pioneer, later plant owners, or society to pay for this often expensive process. But if you criticize the original decision to build a plant on the basis of today's set of heuristics, I most certainly do not concur. As others have said, we must judge the past by its own rule book, not by ours.

These two examples show how easy it is to forget the factor of time in engineering design through inattention. But now let us consider a more troublesome reason why the engineer is often not judged on the basis of good engineering practice at the time of his design. The problem is agreeing on what sota is to be taken as representative of good engineering practice in a specific case. All engineers cannot be asked for their opinions; that is, the sota of the engineering profession cannot be used as a standard. The only recourse is to rely on a panel of qualified experts to give its opinion. But how is such a panel to be constituted? Is membership to be based on age, reputation, or experience? In determining the set of heuristics to represent a sota of a chemical plant, should foreign engineers be consulted? Finally, when best engineering practice is used as a basis for how-safe-is-safe enough for a nuclear reactor, should members of environmentalist groups be included? No absolute answers can be given. But the engineer has never been put off by a lack of information and is willing to choose the needed experts—heuristically. Like any other sota, the set of heuristics used to choose the panel will vary in time and must represent good engineering practice at the time the panel is constituted.

Agreement about the sota in engineering design today is hard enough, but agreement about the set of heuristics appropriate fifty years ago is even harder. Many of the designers of engineering projects still in use are no longer living. Was the steel in the Eiffel Tower consistent with the best engineering practice of its day? With no official contemporary record to document good engineering judgment, history easily erases the engineering profession's memory as to what was the appropriate sota for use in the past. Given the recent rash of product liability claims against the engineer, what is now needed is an archival sota to allow effective implementation of the Rule of Judgment.

RELATIVE TECHNOLOGICAL DEVELOPMENT OF COUNTRIES

The sota in the engineering profession at a given time in a given country is obviously a measure of that time's and that place's technological development and ability to solve technological problems. As such, it will serve as a third example of the importance of the technical word *sota*. A country without access to the sota in engineering practice is at a definite disadvantage. Underdeveloped countries are underdeveloped precisely for this reason. Even among technologically advanced countries, subtle differences in sota result in significantly different products. Competent nuclear engineers have recently reported that a wide variation in testing philosophy in the design of the so-called fast nuclear reactor is evident among major nuclear powers. They report that Americans do extensive testing of design variations and actual components before building the reactor itself; that the French do extensive testing on the full-scale reactor (with the British doing significantly less); and that the Russians prefer to build the reactor first and then see if it can be made to work.

I will refer to the difference between the American and Russian philosophies later and give a name to this heuristic. For now, it is sufficient to recognize that different countries use different sotas when it comes to the testing philosophy of fast nuclear reactors, and that testing philosophy inevitably affects the final product. A colleague once told me he was absolutely convinced that an American engineer was the first to step on the moon because the National Aeronautics and Space Administration required more attention to quality control of individual components than did its competitors. Whether this is true or only the exaggeration of yet another engineer carried away by nationalistic zeal, the country with the most effective heuristics is clearly the most advanced technologically and the best able to respond to new technological challenges. What is needed is research to determine if the sota in engineering practice in America is consistent with the sota in engineering practice worldwide.

As an addendum, I cannot help but wonder if, someday, American engineers who typically speak only English and base their designs only on the heuristics encoded in English will not find themselves seriously behind their multilingual competitors.

ENGINEERING EDUCATION

In engineering education, the lack of a large overlap in the sotas of the average engineering student, engineering professor, and practicing engineer may serve as a fourth example of the sota as a collection of heuristics. Presumably the goal of engineering educa-

tion is to produce a practicing engineer who will perform satisfactorily. Operationally, this goal implies a change in the sota of
incoming freshmen to one that overlaps the sota of the engineer in
the field. This change is difficult to achieve for two reasons. First,
the environment that shapes the sota of the engineer and the one
that shapes the sota of the student are different. The cost of failure
for a practicing engineer can be quite high; the cost for a student is
intentionally limited. In addition, a real design problem may take
years to complete, and it may have a large budget, while the student
is usually limited to a one-semester design course with no budget at
all. Of necessity, the engineer and student work in different environments, and their sotas will evolve differently. Second, the sota of an
engineering professor is not the same as that of a practicing engineer. Often a professor has never solved a real engineering problem
and has little notion of how this should be done. He is therefore
reduced to teaching the theoretical formulas used in design instead
of engineering design itself. Not unexpectly, the result of these two
factors is a noticeable difference in the sotas of the graduating
senior and the practicing engineer.

Engineering educators have had to develop heuristics to deal
with these problems. The traditional approach is to encourage the
practicing engineer to participate in engineering education as a
guest lecturer and to encourage the professor to take a sabbatical
year or consult in industry. Some colleges have also developed
design courses that require students to solve authentic problems
generated by industry, and others have encouraged students to
alternate their formal study with work periods in a cooperative
arrangement with industry. All these remedies have merit, but
focusing attention on the specific set of heuristics that graduating
engineers wrap in their diplomas suggests another approach to
increase the intersection.

The sota of graduating students must contain heuristics that
allow them to efficiently increase the intersection *after* graduation.
While in school students must learn that an engineering design is
defined by its resources and, once in industry, must be alert to the
heuristic used in resource management. They must also realize that
engineering requires that decisions be made amid uncertainty and
look for the heuristics the practicing engineer uses to control the
risk resulting from this lack of knowledge. Engineering education
must not limit itself to trying to achieve an overlap in the sotas of
student and engineer at graduation, but must also teach the novice
engineer to quickly absorb those heuristics that cannot be taught in
school once he is in the industrial environment. The concept of a

set of engineering heuristics, or sota, allows the engineering educator to define the goals of modern education and to develop strategies to achieve them.

THE ENGINEER AND SOCIETY

The relationship between the engineer and society is the last, and most extensive, example of the use of various sotas. It is also one of the most important.

All heuristics are not engineering heuristics, and all sotas are not engineering sotas. As has been observed by other authors, aphorisms, which have all the signatures of a heuristic, are society's rules of thumb for successful living. Too many cooks do not always guarantee that the broth will be spoiled. And what are we to make of the conflicting advice, "Look before you leap" and "He who hesitates is lost"? As with conflcting heuristics, other rules of thumb in the total context select the appropriate aphorism for use in a specific case. These pithy statements are also dated. Recently the sayings, "There's no free lunch," "Everything is connected to everything," and "If you're not part of the solution, you're part of the problem," have appeared. After a decade and a half, the author of "Never trust anyone over thirty" is publicly having second thoughts about his contribution. Aphorisms are social heuristics that encapsulate human experience to aid in the uncertain business of life.

Society solves problems, society uses heuristics, society has a sota. Some of the heuristics used by the engineer and nonengineer are the same, but each reserves some for exclusive use. Few engineers use a Ouija board, astrology, or the *I Ching* in their work, but some members of society evidently do. On the other hand, no layperson uses the Colburn relation to calculate heat transfer coefficients, but some engineers most certainly do. Therefore, the sota of society and the sota of engineers are not the same, but they will have an intersecution, as shown in figure 3.

Here, the stippled sota of engineering has been combined with a crosshatched one representing society. A heuristic earns admission to the small rectangle of intersection by being a heuristic known to both the engineer and the nonengineer. Figure 3 also includes six solid circles labeled 1 through 6 to indicate subsets of heuristics needed to solve specific kinds of problems. Problem 1, lying outside

of the sota of society, requires only engineering heuristics; problem
2 requires some heuristics unique to the engineer combined with
some from the overlap, and so on. As before, both engineering and
society sotas are composites of overlapping individual sotas; therefore, the problems represented by the subsets 1 through 6 will often
require a team effort, possibly including both engineer and nonengineer. Each of these problem areas will now be considered.

Problem 1 requires only engineering heuristics for its solution.
Since the engineer has traditionally responded to the needs of
society, few engineering problems lie in this area.

Problems from area 2 are endemic on the engineer's drafting
board. They require information from society to define the problem
and heuristics exclusively known to the engineer to solve them.
Nonengineers can never completely understand the trade-offs necessary for the design of an automobile. They must delegate responsibility to the engineer to act in their stead and then trust the
engineer's judgment. The alternative approach of restricting the
engineer to problems that require no specialized knowledge—that
is, those requiring no complicated computer models, no advanced
mathematics, and no difficult empirical correlations—would soon

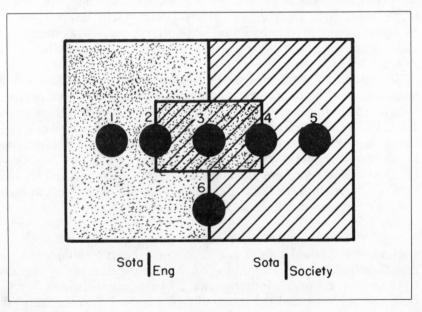

Figure 3

bring the machinery of engineering to a grinding halt. Problems such as number 2 require a joint effort in defining goals and solution strategies, but they also require for their solution heuristics unique to the trained engineer.

Some argue that we are witnessing a shrinking of this area as society disciplines the engineering profession because of disagreement over past solutions. No longer is it sufficient for an engineer to assert that a mass transportation system or nuclear reactor is needed and safe. For a problem such as number 2, confidence in the engineer's judgment outside of the area of overlap is based, in part, on an evaluation of the engineer's performance within the area of overlap. Such an evaluation depends, in turn, on society's understanding of the engineering method. A simple test is in order. Ask the next nonengineer you meet: What does *best* mean to an engineer? How is it related to optimization theory? What is the state of the art? What is technical feasibility? Those unable to respond satisfactorily to these questions are technologically illiterate and in no position to delegate important aspects of their life to the engineer or, more important, to discipline the engineer if they do not agree with the engineer's proposed solution. Given the large number of problems in region 2 and their importance, can society afford humanists who have not even a superficial knowledge of the major ideas that permeate engineering? What is most urgently needed is research to determine the minimum overlap necessary for a nonengineer to be technologically literate.

Problem 3 is simpler to explain. In this area of complete overlap between society and the engineer, the only dispute is over which heuristics are best to define and solve it: those common to both the engineer and nonengineer, those used by the engineer alone, or those used by one of the various subgroups of nonengineers. The politician, economist, behavioral psychologist, artist, theologian, and engineer often emphasize different aspects of a problem and suggest different approaches to its solution. Nothing could be more different than the heuristics *prayer, positive reinforcement,* and *Freudian psychology* when it comes to rearing a child, or than the heuristics used by the politician, economist and engineer when it comes to reducing hunger in the world. Each person speaks of a problem in the accent of his sota. Many options are available in area 3; society's sota must contain effective heuristics to arbitrate between them.

Problem 4 is distinguished from previous ones in that some of the heuristics needed for an acceptable solution are not found within the sota usually attributed to the engineer in his role as engineer.

Remember the San Francisco Embarcadero. Too expensive to tear down, it stands as a monument to a pure engineering solution that failed because the sota used by the engineers did not contain all the heuristics that were important to society. Numerous studies show that compared with the population as a whole, the American engineer is less well read, a better family member, more conservative politically, more oriented to the use of numbers than to general philosophical positions in making a decision, and more goal-oriented. These characteristics may change in the future, of course, and are surely different in different cultures, but the conclusion is inescapable. An engineer is not an average person. Accordingly, when the engineer chooses the important aspects of a problem and assigns their relative importance, at times this model will not adequately represent society. The engineer may think it obvious that there is an energy shortage and that we need nuclear power; some members of society do not agree. The engineer may take it as fact that the scientific view of the world is true; some theologians do not agree.

Only two ways of solving problems in area 4 are possible. Either the engineers can delegate responsibility for certain aspects of a design to laypeople and accept their input no matter how unreasonable it seems, or they can increase their general sensitivity to the hopes and dreams of the human species—that is, increase the overlap of their sotas with that of society. But sensitizing people to the human condition is the responsibility of the novelist, psychologist, artist, sociologist, and historian—in short, the humanist and the social scientist. Another test is in order. Ask the next engineer you meet: What is the central thesis of behaviorism? What is the difference between a Greek and Shakespearean tragedy? An engineer unable to respond satisfactorily to these simple questions is illiterate in the liberal arts and in no position either to delegate important aspects of his or her life to the humanist or, more important, to discipline the humanist if he or she does not agree with the humanist's proposed solution. Given the large number of problems in region 4, and their importance, how can society afford engineers who lack even a superficial knowledge of the major ideas that permeate the liberal arts? What is most urgently needed is research to determine the minimum overlap necessary for an engineer to be liberally educated.

Problem 5 requires heuristics completely outside the expertise of the engineer and is beyond the scope of this discussion.

Finally, problem 6 is included to give equal time to an aberrant view of engineering. Some people, including some engineers, believe that no overlap should exist between the sotas of society and the engineer. In this view, the duty of society is to pose the problems it wants solved, and the duty of the engineer is to solve them using the best techniques available. This view fails because problems evidently exist that cannot even be defined by society without knowing the range of the technically feasible and because solutions evidently exist that cannot be found without knowing a society's value system. I therefore believe that few, if any, examples of problem 6 exist, or if they exist, that they would be solvable. In spite of its obvious flaws, this limited, technical view of engineering is not as rare as it should be.

Figure 3 underscores the effectiveness of the engineer's concept of a sota in the analysis of the relationship between the engineer and society. It must complete the examples intended to show the value of a sota as a tool for bringing understanding to important aspects of the engineering world. With the heuristic, the engineering method, and the sota defined, the discussion returns to one of the major goals of this discussion, to state the rule for implementing the engineering method.

THE PRINCIPAL RULE OF THE ENGINEERING METHOD

Defining a method does not tell how it is to be used. The next task is to find a rule to implement the engineering method. Since every specific implementation of the engineering method is completely defined by the heuristic it uses, this quest is reduced to finding a heuristic that will tell the individual engineer what to do and when to do it. Remembering that everything in engineering is heuristic, no matter how convincingly it may appear otherwise, I have found a sufficient number to implement the engineering method with only one, provided that I make a firm and unalterable resolution not to violate it even in a single instance.

My Rule of Engineering is in every instance to choose the heuristic for use from what my personal sota takes to be the engineering sota at the time I am required to choose.

Careful consideration of this rule shows that engineers normalize their actions against their personal perceptions of what constitutes engineering's best world instead of against an absolute or an eternal or a necessary reality. Engineers do what they believe to be most appropriate measured against this norm. In addition to implementing the engineering method, this Rule of Engineering determines

the minimum subset of heuristics needed to define the engineer.
Earlier in the discussion, the sotas of three engineers, *A, B,* and *N,*
overlapped in a small rectangular subset that included the
heuristics they shared. If, instead of three engineers, all engineers
in all cultures and all ages are considered, the overlap would contain those heuristics absolutely essential to define a person as an
engineer.

This intersection will contain only one heuristic, and this
heuristic is the rule just given for implementing the engineering
method. While the overlap of all modern engineers' sotas would
probably include mathematics and thermodynamics, the sotas of
the earliest engineers and craftsmen did not. While the sotas of
some primitive swordsmiths included the heuristic that a sword
should be plunged through the belly of a slave to complete its
fabrication, the sotas of modern sword manufacturers do not. The
Rule of Engineering is: do what you think represents best practice
at the time you must decide. Only this rule must be present. As long
as the rule is adhered to, neither the engineering method nor its
implementation prejudices what the sota of an individual must
contain for him to be called an engineer.

The goal of this part of my discussion has been to describe how
the engineer responds when encountering a situation that calls for
an engineer. If you desire change; if this change is to be the best
available; if the situation is complex and poorly understood; and if
the solution is constrained by limited resources, then you too are in
the presence of an engineering problem. (What person has not been
in this situation?) If you cause this change by using the heuristics
that you think represent the best available, then you too are an
engineer. What alternative is there? To be human is to be an
engineer.

The definition of the engineering method depends on the
heuristic; the rule of the method and the Rule of Judgment are
heuristics; and the engineer is defined by a heuristic—all engineering is heuristic.

SOME SAMPLE HEURISTICS

1. The engineering method is the use of engineering heuristics.
2. The engineering method is the use of heuristics to cause the
   best change in a poorly understood situation within the
   available resources.

3. Engineering is a problem-solving, goal-directed, and needs-fulfillment activity.
4. Engineering is trial and error.
5. Apply science when appropriate.
6. At some point in a project, freeze the design.
7. Allocate resources as long as the cost of not knowing exceeds the cost of finding out.
8. Allocate sufficient resources to the weak link.
9. Use feedback to stabilize engineering design.
10. Always give yourself a chance to retreat.
11. Make small changes in the state of the art.
12. Work at the margin of solvable problems.
13. Always give an answer.
14. The yield strength of a material is equal to a 0.02 percent offset on the stress-strain curve.
15. One gram of uranium gives one megawatt day of energy.
16. Air has an ambient temperature of 20° centrigrade and a composition of 80 percent nitrogen and 20 percent oxygen.
17. A properly designed bolt should have at least one and one-half turns in the thread.

CONCLUSION

My discussion here has revolved around the engineer and what he does. Engineering, I maintain, is the use of engineering heuristics. With this connection made, the engineering profession should no longer be obsessed with its artifacts; it should concentrate on its art. Enormous potential is unleashed as engineers enter their maturity and use heuristics—their method—to advance engineering heuristics in the future. But more, the responsibility of each *human as engineer* becomes clear. Everyone in society should develop, learn, discover, create, and invent the most effective and beneficial heuristics. In the end, the engineering method is related in fundamental ways to human problem solving at its best.

A Historical Definition of Engineering

EDWIN T. LAYTON, JR.

It is easy to define engineering. Thomas Tredgold (1788–1829) provided one of the earliest and most widely used definitions. Most subsequent definitions, at least those by English speaking engineers, have followed his. To Tredgold and subsequent generations, engineering is "the art of directing the great sources of power in nature for the use and convenience of man."[1]

Tredgold's definition has been widely accepted and repeated with minor modifications by several generations of American engineers.[2] Tredgold was correct in stressing engineering as an art form and the role of engineering in meeting human needs. But I believe that there are major problems with Tredgold's definition. Tredgold's definition, and the innumerable variants upon it that have appeared over more than a century, all treat engineering as a socially neutral, value-free instrument. It is significant that while American engineers accepted Tredgold's definition, they came to differ with him on the social neutrality of engineering.

Whether "use and convenience of man" adequately describes the goal of technology (taken as synonymous with "engineering" for this paper) was subject to some qualifications by American engineers at about the beginning of the twentieth century. The adverse social effects of technology have had a large influence upon engineers, one little noticed outside the profession. Some of the worst horrors of the Industrial Revolution, after all, took place in factories, mines, and mills that were useful and convenient in some sense. Trusts and monopolies similarly might be considered useful and convenient for some, but most Americans regarded them as evil. Beginning in the late nineteenth century, American engineers were increasingly beset with a feeling of guilt and anxiety about the negative social effects of technology—for whose accomplishments, up to that time, they had been taking full credit. Their response was to attempt to define engineering as acting for the good of society.[3]

American engineers reacted to the adverse effects of technology by insisting that its true end is the good of humanity. This involved redefining the goals of engineering. Tredgold's "use and convenience of man" no longer seemed strong enough, and many engineers sought to raise the sights of engineering to a stronger notion of social benefit and moral improvement. Thus George H. Babcock in 1887 thought engineers held a commission from God "to elevate and ennoble man," because "it is the end for which the engineer is commissioned."[4] Charles Hermany in 1904 held that the goal of engineering is to move toward "the brotherhood of man."[5] Though characterizations varied, the new consensus was that the goal of engineering is the good of mankind, not merely use and convenience. Rather paradoxically, conceptualizing the ends of engineering in terms of values has not undermined a belief common among engineers that they and their profession are neutral and value-free, reflecting the objectivity of science.[6]

One of the reasons for the popularity among engineers of Tredgold's definition is his characterization of their role as that of "directing" nature. This role became a very popular part of the ideology of engineering; it made engineers the directors not merely of technology but of social progress as well. But the claim to direct nature for human benefit has never been a literal reality for engineers. Engineers carry out the goals determined by the large corporate entities that control and employ engineering. They "direct" only in the sense that middle-management bureaucrats plan how to carry out orders and supervise subordinates in implementing the goals laid down by top management. As H. B. Gear, a former associate of Samuel Insull, argued, engineers best serve society as "implements of management."[7]

Tredgold's definition reflected an ideology of engineering. It portrayed engineers as agents of progress. But since the advance of science was assumed to be the cause of progress, Tredgold had to claim that engineering was applied science. And Tredgold indeed did link engineering to both applied science and to progress through science. Tredgold, having defined engineering, went on to emphasize engineering as applied science. Here is his fuller definition:

> . . . the art of directing the great source of power in nature for the use and convenience of man; being that practical application of the most important principles of natural philosophy which has, in a considerable

degree, realized the anticipations of Bacon, and changed the aspect and state of affairs in the whole world.[8]

After enumerating some of the specific tasks of engineers, Tredgold concluded by linking engineering as applied science with social progress, which he saw as driven by science:

> This is, however, only a brief sketch of the objects of Civil Engineering, the real extent to which it may be applied is limited only by the progress of science; its scope and utility will be increased with every discovery in philosophy, and its resources with every invention in mechanical or chemical art, since its bounds are unlimited, and equally so must be the researches of its professors.[9]

The theory that engineering, and technology generally, are no more than applied science, in its extreme form, holds that technology consists of applying the fruits of the basic sciences in a more or less mechanical, deductive manner; one can read Tredgold's definition in this light—though he hedged his bet a bit by his references to art and invention. Engineering was imagined to be like a giant machine that received its intellectual fuel from the basic sciences. Engineering sciences were treated as mere applications of the supposedly more fundamental basic sciences. Engineering was therefore hierarchically subordinate to the basic sciences, which provided the intellectual substance for technology.

Tredgold's emphasis upon engineering as applied science is striking and presents a fundamental issue for understanding the historical meaning of engineering. Many definitions of engineering by later American engineers were even more insistent on the theory that engineering is applied science and that its progress depends upon the progress of the basic sciences. This is a problem for the historian because, as a matter of fact, engineering is not just applied science; and its progress is not totally dependent upon prior progress in the basic sciences. To resolve the problem it is enough to recognize that such claims are ideological in nature and not to be taken as literal descriptions. The significance of these claims can be clarified by reference to a few examples.

Tredgold's characterization of engineering as applied science was echoed by Charles Hermany, who claimed in 1904 that "the science of mechanics is the philosophy of engineering. . . . In the study and practical application of mechanics, mathematics is the philosopher's stone."[10] Hermany's statements were paradoxical in that engineering educators, well into the 1920s, were questioning the utility of differential equations and even the calculus.[11] Something of Hermany's motivation is indicated by his linking of engineering to science so that he could argue that engineers are the agents of progress. Hermany thought that engineering is the "herald that brings joy to the multitudes . . .[;] it is their redeemer from despairing drudgery and burdensome labor."[12]

Engineers were eager to claim the social role of the architects of progress, and progress was linked to the advance of science. However, other engineers made significant qualifications that move away from Tredgold's definition. A prominent civil engineer, Thomas C. Clarke, held that "science is the discovery and classification of the laws of nature. Engineering, in the broadest sense, is the practical application of such discovered laws."[13] But this "broadest sense" included engineering itself using scientific methods to gain knowledge. Indeed, Clarke concluded that "engineering is the great creative science."[14] The "applied science" identification also has status implications that were very congenial to engineers. As Clarke went on to note, "When science . . . and experience . . . are united in the same person, then we may truly say that we have seen the evolution of engineering from a craft to a profession."[15]

The relation between the ideological portrait of engineering and engineers as agents of progress through science and attempts to raise the status of engineers was a theme often dwelt on by American engineers in the early twentieth century. In 1895, George S. Morison, a leading civil engineer, sought to demonstrate that engineers are the source of all human progress. In common with many other engineers, he appropriated all technology as the work of engineers. He pictured progress as the simultaneous fruit of inventions and science, which were not seen as distinct activities. Thus Morison could claim, with a certain hyperbole:

> We are the priests of material development, of the work which enables other men to enjoy the fruits of the great sources of power in Nature, and of the power of mind over matter. We are the priests of the new epoch, without superstitions.[16]

In a similarly grandiloquent style, Farley Osgood, a leading electrical engineer, claimed in 1925 that, "The engineers have made the world what it is today; have brought to it industrial progress and economy; have given it the living comforts and, more than all, have shaped the very scheme of living of its inhabitants since civilization began."[17] Significantly, Osgood and Morison also linked the engineers' allegedly exalted social role to the high social status and leadership of society that they thought was coming to the engineer.[18]

The scientific community of the time and the new large corporations also found the applied-science theory of technology ideologically appealing. Basic science deserved funding because of the technological progress it produced. Manufacturers were no doubt pleased with a theory that held that the faults of the particular products that they were foisting upon the public lay not with themselves but were built into the very structure of the universe itself. This ideology found a classic expression in the well-known motto of the 1933 Century of Progress Exhibition: "Science Finds, Industry Applies, Man Conforms."[19]

The applied-science theory of technology was never really developed as a consistent model to explain the history and functioning of engineering. It was purely an ideological construct, and relied on a few examples for its sustenance.[20] It has been replaced in recent years by what is sometimes called the "interactive model." According to this view, engineering disciplines and the basic sciences represent distinct, autonomous, interacting subcommunities—or rather, complex constellations of subcommunities—linked by shared values. These communities are coequal; each is creative and generates knowledge and designs artifacts; and each community may benefit from the work of the other.[21]

Perhaps the most radical challenge to Tredgold's definition of engineering was that of Martin Heidegger, an existential philosopher, who denied that instrumentality of any sort is the fundamental nature of technology. The starting point of Heidegger's philosophy of technology was his denial of instrumentality as an adequate characterization of technology. In his terms, instrumentality is not the "essence" of engineering. Heidegger was led by a purely deductive argument to conclude that the essence of technology lies in "revealing."[22]

I do not find Heidegger's existential philosophy congenial. But it contains important and useful—if not clear—insights even for empirically minded historians. Tredgold's definition had emphasized the instrumental role of engineering, and certainly technology is an instrument for achieving certain types of goals. But it is much more than a mere instrument. Engineering is also a community of people who are practitioners of a creative form of cognition, engineering design, akin to that practiced by artists. Engineering also generates new scientific knowledge using methods and institutions akin to those employed in the basic sciences. Engineers are the guardians of a body of knowledge and a set of values associated with their field or fields of professional practice.

Heidegger's "revealing" in several possible meanings is indeed an essential part of any definition of technology. This would apply particularly to the knowledge-generating dimensions of engineering. But I doubt that it is very helpful to attempt to provide simple characterizations of such complex entities as engineering. I doubt that engineering has a single "essence"; a more pluralistic view would be closer to the empirical realities. Nonetheless, knowledge, surely one part of "revealing," is indeed fundamental and it was long neglected.

Having gained a truly fundamental insight, Heidegger proceeded to miss—or possibly ignore—the highly fruitful cognitive processes of engineering that generate engineering sciences and design knowledge. The form of "revealing" that Heidegger thought is most characteristic of modern technology is a sort of "challenging."[23] While I do not claim to know precisely what Heidegger had in mind, the notion of "challenging" can be given a clear historical meaning. We are challenged by the long term, unexpected, and often undesirable effects of rapid technological change in the modern era. The things that challenge us are the erosion of cherished values and institutions, the reduction of humans to mere cogs in giant social machines, environmental pollution, the psychological and moral dangers of modern mass media, adverse changes in the status of women and minorities, and the excessive concentration of power and wealth in a few hands. We are particularly challenged by threats to safety and even the continuation of the human species implicit in modern weapons systems, or by unforeseen dangers to the environment and to humans arising from technological activities. These and other challenges have stimulated much work in the externalist history of technology as well as in other disciplines.

Heidegger is rather vague about the other sorts of "revealing" produced by engineering. Though he appears to have believed that engineering is a source of new knowledge, his approach apparently did not bring him into contact with those sources which would have informed him about the precise nature of engineering knowledge. Thus he accepted without question a critical error that has vitiated much thought about the role of engineering. Heidegger maintained that "modern technology is something incomparably different from all earlier technologies because it is based on modern physics as an exact science."[24] We should not be too harsh on Heidegger here; the applied-science model was still widespread when he wrote, and its advocates could cite eminent engineers such as Tredgold as authorities for their claims.

Heidegger was incorrect in thinking that modern technology is applied physics. Both Tredgold and Heidegger missed the crucial role of engineering as a mode of inquiry into human artifice and natural phenomena parallel to but generally independent of physics and other basic sciences. That is not to say that engineers do not use physics, mathematics, chemistry, and other sciences whenever they can. But the needs of design are rarely satisfied by extant science. To design successfully and efficiently, engineers are compelled to generate the new scientific knowledge needed by engineering designers. To this end, engineers have modified and adapted the methods and institutions employed in the basic sciences to serve engineering design, giving engineering a structure, methods, and values different from though akin to those in the basic sciences. Scientific knowledge has been codified in distinctive "engineering sciences."[25] These sciences might equally be called "the sciences of design," since their role is primarily to facilitate design and only secondarily to enlarge the domain of physical science.[26] Engineering design and engineering science were not explicitly noted by either Tredgold or Heidegger. (To his credit, Tredgold may have used the modifying terms "art" and "invention" in characterizing engineering precisely because engineering involves subjective design as well as objective sciences of various sorts.)

Having accepted the false premise that engineering is applied physics, Heidegger did not, however, draw the usual inference that technology is nothing more than applied science. It is to his credit that Heidegger rejected the theory that technology and engineering are applied science. In this case, his very ignorance may have caused Heidegger to raise questions of considerable interest; it may have forced him to find indirect rather than direct arguments. Heidegger noted that physics is based on technological instrumentation—a point stressed by many historians of technology, and most notably by the late Derek J. Price. If modern science is dependent upon technology in such an essential and vital way, then it makes little sense to claim that science produced technology.

Heidegger was also led to make a more abstract argument. He held that technology created the intellectual atmosphere out of which physics arose.[27] Heidegger's argument is very interesting and it too can be interpreted in empirical historical terms. At a gross level, one can note that in the centuries prior to the scientific revolution engineers and architects such as Leonardo, Alberti, and

Brunelleschi—and even some medieval technologists—developed methods of experimentation with machines and developed mathematical theories for technology that did provide important foundations upon which Galileo and other pioneers of the scientific revolution of the seventeenth century were to build.[28]

Heidegger's rejection of instrumentality was fruitful, and it could lead to important historical realities—despite the fact that Heidegger's "revealing" and "challenging" are not so different from the instrumentality that he rejects. The revelations and challenges that engineering produces cannot be separated from the normal, instrumental operations of engineering communities. Besides design, engineers test, sell, supervise, produce, and operate engineering artifacts and engineering systems. Many engineers have risen to positions in the management of large private corporations and governmental agencies. In performing their normal functions—for example, in inventing automobiles, weapons systems, and the like, and then advocating their use, that is, in generally functioning as "systems builders"—engineers both act instrumentally *and* create new technologies that "challenge" society. Further, engineering value systems place a greater value on "doing" (that is, instrumentality) than upon "knowing," though both categories of action are valued.

Engineering is now recognized as an important source of scientific and technological knowledge. It has its own characteristic cognitive form. The creative process by which the intellectual substance needed to create a new engineering artifact or system is brought into being is called "engineering design," or simply "design."[29] Engineering design is the synthetic process in which engineering artifacts and systems are first conceptualized and then by degrees brought into being.[30] Clearly, engineering design must play a role in any definition of engineering.

Design, although it has been influenced by science, remains distinct. This is the case particularly in design's focusing upon solving "ill-defined problems," so characteristic of design, rather than on the "well-defined problems" characteristic of many areas of mathematics and physical science. Engineering design is driven explicitly by social needs, and it incorporates human values as an integral part of the design problem, things not usually attributed to basic science. Design is also characteristically iterative and cyclical rather than linear; this reflects the imperfection of knowledge in design problems, so that new knowledge inputs can and usually do appear at any stage of the design process, causing a redesign of earlier stages. Designs are normally "optimized" to better adapt them to the specific needs and values that the design is intended to serve. There is in science no exact analog to optimization—though one may observe that scientific theories usually aim at the greatest generality of knowledge expressed in the most economic way. There is less of a critical priority in basic scientific theories that they be capable of producing specific, usually numerical, results with minimal computational difficulties. Perhaps most critically, design is still synthesis, not analysis, however much it might use science and be influenced by it in achieving engineering synthesis.[31]

Some engineers have sought to define engineering in terms of design. To such people, the true engineer is the designer. Design is therefore the function that most characterizes engineers. For example, in 1895, George S. Morison gave an address on the "true meaning and position of the profession." In this address he argued that the true engineer is the one who understands the underlying principles and who can design a machine or structure, as against someone who can only operate it.[32] Similarly, an electrical engineer, William McClellan, held that engineering combines two distinct traditions, one theoretical or scientific and the other practical or mechanical. McClellan argued that to these traditions there correspond three technical types: the applied scientist, who is limited to science and theory; the mechanic, who is the master of technique; and the designer, who combines the scientific with the practical. To McClellan, the designer is "the real engineer."[33]

The identification of engineering with design, rare in the first half of this century, has become much more widespread since the Second World War. The renewed stress upon design has produced publications with titles such as *Design: The Essence of Engineering*.[34] But if ideology has only recently begun to catch up with reality, engineering practice has always emphasized design as the defining characteristic of engineering. The role of design in defining engineering is particularly clear in the United States, where government has not imposed an arbitrary, legal definition, as it has in most European countries. This has led to a definition of engineering that virtually all American engineers accept: engineering is a "vaguely bounded nucleus within a large body of technical workers."[35] To American engineers, the thing that separates the inner core of professional engineers from other technical workers is the ability to design. American engineering societies have generally accepted "ability to design" as the defining characteristic of the "true" professional engineer. However, since not all engineers are designers, a minor change is necessary; this takes the form of "qualified to design."[36]

It is important, when rejecting the false theory of technology as applied science, not to fall into the complementary fallacy of holding that applied science plays no role in engineering or in design. Not only do engineers use as much basic science as they find useful, but engineering has come increasingly to stress physics, chemistry, mathematics, and other basic sciences. A glance at the professional literature produced since the Second World War indicates the large extent to which engineering subjects have come to be cast in the mold of classical physics and other basic sciences.[37] Moreover, it is a fact that a substantial number of engineers—and particularly university faculty—have become researchers in a cluster of engineering sciences.

As Zehev Tadmor has recently noted, "The engineering faculty have accepted the challenge and responsibility of formulating and organizing the engineering sciences into teachable bodies of knowledge . . . These engineering sciences form the core of the present-day engineering curriculum."[38] Nor should we miss the fact that there are important innovations in which the applied-science model does approximate the facts. That is, there are important technological innovations that have had their origins in discoveries of the basic sciences, are developed by engineering scientists, and finally produce innovations by means of engineering design.[39] Science, in Heidegger's terms, is not the essence of engineering. But science, at several levels, has proven itself to be a very useful tool for engineers, and it has had a profound impact upon engineering practice.

To better show the link between "revealing" and the instrumental or functional role of engineering, it is helpful to consider the social relativity of engineering. An important part of any philosophy of technology, in my opinion, should be the recognition that all engineering artifacts and systems are not socially neutral, that they are permeated by values. All designs are shaped in a fundamental way by values. This fundamental truth was obscured for some time by the applied-science theory of technology.

Heidegger saw values emerging in the social effects of technology in his concept of "enframing," and also in the mode of "revealing" of modern technology though "challenging."[40] But Heidegger did not spell out the means by which values become essential parts of engineering artifacts and systems. There are several avenues through which social values influence design and the artifacts and systems produced by design. These include engineering styles, the social determination of engineering goals, and the need to optimize engineering designs. That is, in Heidegger's terms, the *telos* of engineering is determined by society, and not by scientists or by engineers.[41]

Engineering designs are influenced by engineering style; this reflects the fact that design is still an art and carries with it many tacit individual and cultural assumptions about how to implement any given set of design goals. These assumptions are not entirely objective or scientific. They have changed over time and often differ among different cultures. Engineering and engineering design may be regarded as decision-making processes. It is axiomatic that there are almost always choices for the designer at each step in the design process. From what has been said, it is clear that any criterion for making choices will involve value decisions and will constitute a style of engineering. Different national and cultural values lead to different choices being made and affect the end product.[42]

It is one of the defects of Heidegger's essay that he saw technology producing art—in the sense of fine arts—only as a distant utopian goal.[43] Here, too, Heidegger's method of inquiry has left him apparently ignorant of the empirical realities. While the Industrial Revolution served to diminish the traditionally high value of art in engineering designs, the artist-engineer has, fortunately, never died out. The linkage of engineering with art—though it has, at times, been buried under piles of ugly, sterile artifacts—has never been wholly lost. As David Billington has pointed out, the Industrial Revolution produced its own tradition of structural artists, starting with Thomas Telford, the founder of the Institution of Civil Engineers. Indeed, Telford wrote on the aesthetics of engineering and on the practice of engineering as a fine art.[44] This tradition has continued among a number of engineers, including notably Robert Maillart, down to the present.[45] John Kouwenhoven and John Kasson have discussed the lively controversies over the aesthetics of engineering and technology in nineteenth-century America.[46] Carl Condit is notable for his studies of art and American building.[47] Engineering has never ceased to be a mode of artistic expression, and the restoration of engineering as art is one of the priorities of the next generation if we wish to make our human-made environment livable and pleasing to the senses. One part of any definition of engineering must include the engineer as artist.

The rise of operations research and systems engineering since the Second World War has served to emphasize precisely how values are built into engineering. In particular, systems analysis has thrown light upon optimization. Optimization is one of the important ways in which social values and goals have come to shape engineering. That is, designs are contrived so as to maximize intended benefits and minimize certain undesirable attributes, but not others, by means of "trade-offs." In effect, optimization represents the "fine tuning" that adapts an engineering artifact or system to its intended purposes and to the values imbedded in these goals. Granted the need to optimize designs, there is a need to assign values to the various performance characteristics to be considered in the optimization.

The question of optimization is one that has been critical in the redefinition of engineering by engineers. In the early days of engineering design, the problem was simply to get a working system. Improvements were simply taken for granted as a natural part of the design process. In early textbooks on design, such as that of Jacques Eugène Armengaud, optimization was often confused with efficiency and was seen as a natural product of concentrating upon one's technical task. Armengaud provided rules for maximizing the efficiency of a machine—for example, a hydraulic turbine. But he failed to deal with optimization explicitly. This was consistent with a long-standing tendency of engineers to see design primarily as a scientific problem.[48] The confusion of efficiency and optimization continued well into the twentieth century. Efficiency maximizes output in relation to input; optimization adapts engineering systems to particular goals and values. The optimal design solution may not, in fact, be the most efficient. If safety for humans and protection of the natural environment are given a high priority in optimizing an artifact or system (say, for example, a nuclear power plant), the added costs may well make the system less efficient than one which is less safe.[49]

It is ironic that the search for objective, mathematical methods of optimizing engineering designs has led to this proof that all engineering designs are permeated with values—that they are, in consequence, subjective in a very fundamental sense. Because of this discovery, modern engineers have for the most part abandoned claims that engineering is totally objective and scientific. In place of these earlier formulations, engineers have shifted to a more subjective, value-laden view of the products of design. This change was produced in large part by the new emphasis upon mathematical methods of optimization developed in the post–World War II era as fruits of the development of operations research and systems analysis. Specifically, it became possible to model engineering systems in terms of linear algebra and to perform mathematical operations on the model in order to optimize the system.[50] As Morris Asimow has pointed out: "If we accept the fact that changes in the design parameters will cause concomitant changes in the output variables, there will be one choice among the gamut of satisfactory choices of design parameters which will be as good as or better than any other."[51]

Systems engineers were able to express any given engineering design mathematically in what is called the "criterion function" or "objective function." But to construct this function they found that they had to represent the system as the sum of a number of terms, each consisting of a criterion variable (x) multiplied by a weighing coefficient (a), usually chosen on an arbitrary zero-to-nine scale. The weighing coefficient gives a quantitative measure to the value of the particular criterion. This has the effect of explicitly incorporating values in every design. Indeed, it clearly subordinates technical criteria to value judgments in engineering design.

As Thomas T. Woodson, author of perhaps the most popular contemporary text in engineering design has put it:

> If this expression, $a_i x_i$ is a reasonable symbolic model of the system, at what point do all the other mathematical models of system behavior fit? How about circuit analysis, thermodynamics, feedback control and stability, structural integrity, materials and process analysis? Are all these rigorous mathematical tools to be subjugated to subjective decisions from the top? The answer is, of course, that they are subjugated; but that as tools they do fit in at lower levels of the design process, aiding in producing adequate decision-making information.[52]

Engineers have come to recognize that engineering designs are shaped by a variety of values. Thus Woodson notes:

> As we look behind the scenes, we find major influences on decision-making coming from the individual's own value system, from that of his organization, and from the culture, as well as from the technology.[53]

Tredgold's definition had restricted engineering to the artifacts it designs and builds; he did not suggest any involvement with the social structures used by or influenced by engineering. Modern scholars and modern engineers have come to see engineering and society as part of the system, and not sharply separated.

The new system engineering has had a rather distorted impact upon historical scholarship concerning engineering. Engineering requires social structures for its implementation. Economic historians and historians of technology have come to see the vital relation of technology to society in terms of systems. This has the important advantage of emphasizing the interaction between technology and engineering on the one hand and various social structures on the other. To those who think of technology and society as interacting systems, there can be no technology isolated from society or society isolated from technology. Questions of the "impact" of technology on society, from the perspective of an interactive, ongoing process, are one-sided and misleading.

For a technology to be implemented, appropriate social frameworks need to be created or adapted from existing social structures. It is, of course, for this reason that technology has been a principal cause of social change. The social change and dislocations associated with rapid technological change have been particularly evident since the Industrial Revolution. The social framework created to implement technology then can constrain human behavior in a variety of ways, many of them undesirable. Thus it is somewhat unrealistic to talk about the social effects of technology, implicitly separating engineering from its social interactions.

The systems approach gives us another definition of the engineer and of engineering: the engineer is a systems builder, and engineering is the art and science of building systems. While the systems approach has served to emphasize the social relativity of technology and engineering, historians who have used the systems concept have not attempted to achieve the conceptual rigor of mathematical systems analysis as it is employed by engineers. In particular, those who have discussed systems often present the particular social arrangements constructed to implement a technology as if they were objective social necessities. Thus the system approach has, in some cases, become a sort of defense of past decisions and injustices. If the interrelated systems are seen as objective necessities, then the entrepreneur is seemingly justified, no matter how selfish, antisocial, or evil the acts may have appeared to contemporaries. The systems approach to economic and technological history, therefore, may become a means of legitimating the misdeeds of the systems builder. In effect, such historians ignore optimization and the trade-offs between the various goals involved in a complex sociotechnical decision-making problem. Social systems are surely not less subjective nor less influenced by values and interests than are engineering systems.

If optimization were added to the historical and economic analysis of the social systems associated with technology, then a number of issues would have to be faced. Automobiles, for example, were optimized for maximum profit to the manufacturer rather than for service to the public. In particular, the lack of concern for safety, fuel economy, and the environment in automobile design says something about the relative weight given to the various criteria involved in designing an automobile.[54]

Heidegger discusses the creation of social structures to implement technology under the title "enframing." Much of his discussion is enlightening. But he is, I believe, incorrect in assuming that this is a new phenomenon associated with modern technology, and not apparent in earlier craft-based technologies. He is concerned with the ways that technological systems can constrain humans, but less sensitive to the ways that they have liberated humanity in a number of areas, including access to education, culture, and information generally. Records, tapes, and compact disks contribute to noise pollution; they are also the means of making classical music and great performers available to a mass audience.[55]

Though social change is an inseparable part of engineering, it is, practically speaking, a topic that requires methodologies that go beyond history. The literature is vast, and I will therefore conclude my discussion of the historical definition of engineering with an admonition to interpreters: engineering involves social change in an essential way. It is ironic that engineers, archetypical social conformists and "solid citizens," should have produced more drastic and more lasting social changes than all of the avowed social revolutionaries put together. Thus, my final historical definition of engineering is this: it is an instrument of social change and social revolution.

Such a claim superficially sounds like the discredited claims about engineers as architects of social progress. Technology—and science—can transcend the organizations that fund and control them. In this sense they may have a profound influence independent of their cultural captivity to corporate organizations. At least two phenomena are involved. The course of research in both science and engineering may lead to unforeseen discoveries. And, as William F. Ogburn's well-known theory of social lag has made apparent, technologies can do two things that go beyond the intentions of those who promote or market them. First, they can produce social disequilibrium, as the technology-society interaction produces social problems more rapidly than society can find mechanisms of accommodation. Secondly, technologies have long-range effects that are not foreseen, or even foreseeable in many cases, by the agencies producing them. There is therefore an irony. The claims made by engineers that they shape society have been validated in terms of secondary or long range effects of engineering that are undesirable or harmful. Engineering is a "sorcerer's apprentice" much more than it is the leader of humanity into a new utopia.

NOTES

1. Thomas Tredgold, quoted in Charles Hutton Gregory, "Address of the President," *Institution of Civil Engineers, Minutes of Proceedings* 27 (January 1868): 181–82. This definition has an "official" character lacking in most speeches by engineers. The Council of the Institution of Civil Engineers (founded in 1818) invited Tredgold to define the objects of the institution; his reply, a letter written to the secretary on 4 January 1828, contained the definition quoted above. The purpose of the inquiry had been to incorporate his definition in the application of the institution for a charter from the government. Tredgold's definition was of civil engineering. At that time, civil engineers still claimed that the term meant all nonmilitary engineers. This insularity was continued in America by the American Society of Civil Engineers, which claimed to represent all engineers, even though mechanical and other specialized branches of engineering were extant and organized. American civil engineers did not give up the claim to represent all engineers until after 1900.

2. For examples of the affirmation of Tredgold's definition by American engineers, see James R. Croes, "A Century of Civil Engineering," *Transactions of the American Society of Civil Engineers* 45 (June 1901): 599–616. For noncivil engineers see Frederick R. Hatton, "The Mechanical Engineer and the Function of the Engineering Society," *Proceedings of the American Society of Mechanical Engineers* 29, pt. 2 (December 1907): 597–98, and M. L. Holman, "The Conservation Idea as Applied to the American Society of Mechanical Engineers," *Proceedings of the American Society of Mechanical Engineers* 31 (January 1909): 4.

3. Edwin T. Layton, Jr., *The Revolt of the Engineers: Social Responsibility and the American Engineering Profession,* 2d ed. (Baltimore: Johns Hopkins University Press, 1986), pp. 58–74 and passim.

4. George H. Babcock, "The Engineer, His Commission and His Achievements," *Transactions of the American Society of Mechanical Engineers* 9 (1887–88): 23–37.

5. Charles Hermany, "Address," *Transactions of the American Society of Mechanical Engineers* 53 (December 1904): 462.

6. This makes sense if one notes the fact that engineers do not, in fact, call the shots. That is, their work may be seen as an objective implementation of goals and values handed down by employers and implemented objectively by engineers. These various notions are part of an ideology of engineering, however, and it is not particularly helpful to take such ideologically based statements literally. Part of this same ideology was the self-image of the engineer as a superhuman logical thinker; such statements are often coupled with highly illogical or biased statements. (See Layton, *Revolt of the Engineers,* pp. 58–74.)

7. H. B. Gear, "Engineering as an Implement of Management," *Electrical Engineering* 61 (August 1942): 426–27. On the subordination of engineering to business, see also Layton, *The Revolt of the Engineers,* pp. 1–19 and passim. This theme has been addressed by others, notably Steven L. Goldman—for example, in "The Social Captivity of Engineering," this volume, pp. 121–45.

8. Tredgold, "Address of the President," p. 182.

9. Ibid.

10. Hermany, "Address," p. 459.

11. H. D. Gaylord, "The Relations of Mathematical Training to the Engineering Profession," *Engineering Education* 7 (October 1916): 54–72.

12. Ibid., p. 464.

13. Thomas C. Clarke,"Science and Engineering," *Transactions of the American Society of Civil Engineers* 35 (July 1896): 508.

14. Ibid., p. 518. Many engineers considered design to be scientific, so both design and engineering science may be included in Clarke's definition.

15. Ibid., p. 519

16. George S. Morison, "Address," *Transactions of the American Society of Civil Engineers* 33 (June 1895): 467.

17. Farley Osgood, "The Engineer and Civilization," *Journal of the American Institute of Electrical Engineers* 44 (July 1925): 705.

18. Ibid., pp. 706–7.

19. John Staudenmaier, "What SHOT Hath Wrought and What SHOT Hath Not: Reflections on Twenty-five Years of the History of Technology," *Technology and Culture* 25 (October 1984): 709–10. See also Lowell Tozer, "A Century of Progress, 1833–1933: Technology's Triumph over Man," *American Quarterly* 4 (Spring 1952): 206–9.

20. Most of these have been subsequently refuted by the modern, critical discipline of history of technology. The interpretation of Watt's invention of the separate condenser as an application of Joseph Black's discovery of latent and specific heats is a classic; it has been repeatedly refuted, but it keeps being reborn. In such a case one may conclude that ideology, not historical fact, is the motivation for the repetition of these false and discredited claims. See the refutation of this myth by D. S. L. Cardwell, *From Watt to Clausius: The Rise of Thermodynamics in the Early Industrial Age* (Ithaca: Cornell University Press, 1971), pp. 34–57.

21. For a discussion of the "interactive model" and the literature on the subject, see Edwin T. Layton, Jr., "Through the Looking Glass, or News From Lake Mirror Image," *Technology and Culture* 28 (July 1987): 597–605.

22. Heidegger defined the essence of engineering in the following passage: "We have arrived now at *aletheia,* at revealing. What has the essence of technology to do with revealing? The answer: everything. For every bringing-forth [*poiesis*] is grounded in revealing [*aletheia*]. . . . Technology is a mode of revealing." (Martin Heidegger, "The Question Concerning Technology," in Heidegger, *Basic Writings,* ed. D. Krell [New York: Harper & Row, 1976], p. 294.) Though *poiesis* has a broader meaning than artistic creativity such as is found in poetry and the fine arts, Heidegger also uses *poiesis* to describe this latter form of creativity. (Heidegger, "Question Concerning Technology," pp. 315–16.)

23. Heidegger, "Question Concerning Technology," pp. 296–99.

24. Ibid., p. 295.

25. Edwin T. Layton, Jr., "Mirror Image Twins," *Technology and Culture* 12 (October 1971): 562–80. On science and design in engineering see also Layton, "Technology as Knowledge," *Technology and Culture* 15 (January 1974): 31–41; "American Ideologies of Science and Engineering," *Technology and Culture* 17 (October 1976): 688–700; and "Engineering and Science as Distinctive Activities," in *Technology and Science: Important Distinctions for Liberal Arts Colleges,* Davidson-Sloan Series in the New Liberal Arts, vol. 1 (Davidson, N.C.: Davidson College, 1984), pp. 6–13. On engineering sciences see also Layton,

"Science as a Form of Action: The Role of the Engineering Sciences, *Technology and Culture* 29 (January 1988): 82–97.

26. See Layton, "Through the Looking Glass," pp. 603–7. The engineering sciences are similar in general terms to other sciences of the artificial associated with professions such as medicine, as pointed out by Herbert A. Simon, *The Sciences of the Artificial* (Cambridge: MIT Press, 1969).

27. Heidegger, "Question Concerning Technology," pp. 302–3. My sometime student, Joachim "Joe" Breuer made me aware of the subtlety of Heidegger's argument here.

28. See for example the following: Alexander Keller, "Mathematicians, Mechanics and Experimental Machines in Northern Italy in the Sixteenth Century," in *The Emergence of Science in Western Europe,* ed. M. P. Crossland (London: Butterworth, 1975); and the same author's "Pneumatics, Automata, and the Vacuum in the Work of Giambattista Aleotti," *British Journal for the History of Science* 3 (1967): 338–47. See also Ladislao Reti, "Il Moto dei proietti e del pendolo secondo Leonardo e Galileo," *Le Machine* 1 (1968): 63–89; Thomas Settle, "Ostilio Ricci, a Bridge Between Alberti and Galileo," *XII Congrès D'Histoire des Sciences, Actes* (1968), pp. 121–26; Samuel V. Edgerton, *The Renaissance Rediscovery of Linear Perspective* (New York: Basic Books, 1975); and Vernard Foley, "Leonardo's Contributions to Theoretical Mechanics," *Scientific American* 225 (1985): 108–13. The modern history of technology has shown that there is no sharp break with the Renaissance, and the roots of industrialism and science are linked also with medieval technology, as for example the masons who built the cathedrals. See Robert Mark, *Experiments in Gothic Structure* (Cambridge: MIT Press, 1982), and James S. Ackerman, "Ars Sine Scientia Nihil Est, Gothic Theory of Architecture at the Cathedral of Milan," *Art Bulletin* 26 (1949): 84–111.

29. There are a number of studies that shed light on the roles of science and design in the process of innovation. Many of these studies are biased, but a very useful and revealing study was that of D. G. Marquis and Sumner Myers, *Successful Industrial Innovations,* National Science Foundation Report, NSF 69-17 (Washington, D.C.: NSF, 1969). In the large sample examined by Marquis and Myers only 5 percent of the innovations had their origin in systematic research, and 15 percent of the innovations were in some way facilitated by systematic research, mostly to implement an idea for an invention or novel design (table 19, p. 46).

30. Edwin T. Layton, Jr., "Science and Engineering Design," *Annals of the New York Academy of Sciences* 242 (1984): 173–81. See also David P. Billington, *The Tower and the Bridge* (New York: Basic Books, 1983), pp. 8–10, 41–44, 230–32; and Billington, "Design as Art and Invention," in *Technology and Science: Important Distinctions for Liberal Arts Colleges,* pp. 14–26. A standard textbook is Thomas T. Woodson, *Introduction to Engineering Design* (New York: McGraw-Hill, 1966).

31. For analyses of engineering design as a form of cognition see John M. Carroll and others, "Aspects of Solution Structure in Design Problem Solving," *American Journal of Psychology* 95 (June 1980): 269–84; Carroll and others, "Presentation and Representation in Design Problem Solving," *British Journal of Psychology* 71 (1980): 143–53; and Ashok Malhotra and others, "Cognitive Processes in Design," *International Journal of Man-Machine Studies* 12 (1980): 119–40.

32. George S. Morison, "Address at the Annual Convention," *Transactions of*

the American Society of Civil Engineers 33 (1895): 471–72. Morison argued that the true engineer is a civil engineer, that civil engineers comprise all nonmilitary engineers and are not just a subbranch of engineering. This position was already obsolete when Morison enunciated it. It had to do with internal politics in connection with efforts to unify the engineering profession in America.

33. William McClellan, "A Suggestion for the Engineering Profession," *Transactions of the American Institute of Electrical Engineers* 32, pt. 2 (1913): 1271–72.

34. M. W. Lifon and M. B. Kline, *Design: The Essence of Engineering* (Los Angeles: U.C.L.A., Department of Engineering publication EDP 5–68, 1968).

35. William E. Wickenden, "Engineering Education Needs a 'Second Mile,'" *Electrical Engineering* 54 (May 1935): 472. Quoted in Layton, *The Revolt of the Engineers,* p. 26.

36. It is significant that the ability to design has been accepted as the primary thing that makes an engineer, even when the actual standard has been lower. For example, the American Society of Mechanical Engineers had a de facto primary qualification of being in "responsible charge" of engineering work, but in the constitution adopted in 1904 the requirement was phrased as, "A member must have been so connected with engineering as to be competent, as a designer or as a constructor, to take responsible charge of work in his branch of engineering." (See "American Society of Mechanical Engineers, Constitution," *Transactions of the American Society of Mechanical Engineers* 25 [1904]: xii. See also Layton, *Revolt of the Engineers,* pp. 26–27, 30, 39, 49n, 51n, 80, 88–89.)

37. Olaf A. Hougen's "Seven Decades of Chemical Engineering" (*Chemical Engineering Progress* 73 [January 1977]: 89–104) is a history of the evolution of the sciences employed by chemical engineers.

38. Zehev Tadmor and others, *Engineering Education 2001* (Haifa, Israel: Neaman Press, 1987), p. 14.

39. Illinois Institute of Technology, *Technology in Retrospect and Critical Events in Science,* 2 vol. (Chicago: Illinois Institute of Technology, 1968). Though flawed in many ways and biased as NSF's reply to its critics inside and outside the government, this study does show examples of important technologies that had their origins in basic science. The NSF funded a follow-up that extended the number of innovations from five to ten. See Battelle Memorial Institute, Columbus Laboratories, *Interactions of Science and Technology in the Innovative Process: Some Case Studies* (Columbus: Battelle Memorial Institute, 1973).

40. Heidegger, "Question Concerning Technology," pp. 296–302. I will return to "enframing" and "challenging" below; though I differ significantly with Heidegger, here too he has touched upon fundamental realities about technology and engineering that are part of their essential nature (and hence of any historical definition of engineering).

41. Heidegger, "Question Concerning Technology," p. 291. Heidegger uses "telos" in the broader sense of "giving bounds" or "completing." He seems to want to exclude the usual translation as "aim" or "purpose." I accept the broader definition but follow the conventional wisdom in including "aims" and "purposes" as legitimate translations. It is, however, helpful to think that "society" (e.g., the paymasters who set engineers at their tasks) determines not merely the aims and purposes of engineering, but also sets bounds to it, and helps to define "completion." Nature does, of course, set boundary conditions for engineering. Engineering cannot repeal the law of gravity. But this constraint has been overstated in the past. The law of gravity was a long-standing challenge for engineers and inventors to circumvent the law and learn to fly. On "telos" see also Don Ihde, *Tecnics and Praxis* (Dordrecht, Holland: Reidel, 1979), pp. 40–50.

42. Eda Kranakis, "Social Determinants of Engineering Practice: A Comparative View of France and America in the Nineteenth Century," *Social Studies of Science* 19, no. 1 (February 1989): 5–70. See also Edwin T. Layton, Jr., "European Origins of the American Engineering Style of the Nineteenth Century," in *Scientific Colonialism*, ed. Nathan Reingold and Marc Rothenberg (Washington, D.C.: Smithsonian Institution Press, 1987), pp. 151–66.

43. Heidegger, "Question Concerning Technology," pp. 315–16. Heidegger says, "Modern technology does not unfold into a bringing-forth in the sense of *poiesis*" (p. 296).

44. Billington, *Tower and the Bridge*, pp. 38–41.

45. David P. Billington, *Robert Maillart's Bridges: The Art of Engineering* (Princeton: Princeton University Press, 1979).

46. John A. Kouwenhoven, *The Arts in Modern American Civilization* (New York: Norton, 1967); and John Kasson, *Civilizing the Machine: Technology and Republican Values in America, 1776–1900* (New York: Penguin, 1976), pp. 139–80.

47. Carl Condit, *American Building* (Chicago: University of Chicago Press, 1968).

48. Optimization is not clear in the earliest design textbooks, though efficiency is. For America, an influential work that brought French descriptive geometry and advances in the science of machines was Jacques Eugène Armengaud and others, *The Practical Draughtman's Book of Industrial Design, Forming a Complete Course of Mechanical Engineering and Architectural Drawing*, rev. W. Johnson (Philadelphia: Henry Cary, 1871). See also Jacques Eugène Armengaud, *Traité theorique et pratique des moteurs hydrauliques* (Paris: A. Morel, 1868).

49. See the discussion on optimization in Woodson, *Introduction to Engineering Design*, pp. 258–64 and passim.

50. The original word was "operations research." There has been a tendency to broaden the name by the use of the term "systems analysis." The two are not really different, and are sometimes used interchangeably. See Wayne L. Winston, *Operations Research: Applications and Algorithms* (Boston: Duxbury Press, 1987), pp. 606–35, which contains a discussion of "goal programming," an extension of linear programming, which emerged in the postwar world to handle complex problems of scheduling, but which has been subsequently expanded to cover a large range of engineering problems, especially in relation to management.

51. Morris Asimow, *Introduction to Design* (Englewood Cliffs, N.J.: Prentice-Hall, 1962), p. 84. Asimow was one of the founders of modern design methods and outlook, which as noted involved a fundamental shift from seeing designs as scientific to seeing them, at base, as social constructs. This latter was an unintended byproduct of mathematical methods of optimization, the intended goal.

52. Woodson, *Introduction to Engineering Design*, p. 205.

53. Ibid., p. 204.

54. For a bibliographical review of the economic historians who have written about technology as systems, see Nathan Rosenberg, *Inside the Black Box: Technology and Economics* (New York: Cambridge University Press, 1982), pp. 59–62. Much recent work has been done on systems by historians of technology. A notable example is Thomas P. Hughes's *Networks of Power: Electrification in Western Society, 1880–1930* (Baltimore: Johns Hopkins University Press, 1983), pp. 1–17 and passim. See also the same author's "Emerging Themes in the History of Technology," *Technology and Culture* 20 (October 1979): 697–711.

55. Heidegger, "Question Concerning Technology," pp. 300–309.

Engineering as Productive Activity: Philosophical Remarks

CARL MITCHAM

According to classic reports, the man who brought philosophy down from the heavens to dwell among human beings and to engage in human affairs also took philosophy into the workshops of the artisans and craftsmen (the *cheirotechnai* and *demiourgoi*), and even found there a respectable, though not wholly philosophical, understanding.[1] Yet the history of philosophy in the West has largely failed to follow this example. Indeed, philosophy has retained a strong tendency to return, in one way or another, to the heavens from whence it came. Even in those post-Renaissance philosophies that emphasize the active character of the human and seek, in different ways, to deal with the real-life world, discussion of an activity that is uniquely characteristic of the modern world, namely engineering, is conspicuous by its absence.

Since Socrates, one fundamental characteristic of philosophical discourse has been an inquiry into the "whatness" or essential nature of things. According to both Plato and Aristotle this whatness is identified with form, which is also the foundation for and thereby able to be grasped through its distinctive operations or functions. Operations or activities, like forms, can thus be described in terms of genus and species (even though in the strict Aristotelian sense actions, not being substances, do not have essences). The hypothesis here is that the genus for understanding engineering is human activity—that is, that engineering is to be defined not so much in terms of special types of objects or of certain kinds of knowledge or of unique volitional commitments[2] as in terms of a particular species of behavior. However, given the extreme diversity of human activities—from speaking, playing, politicking, and praying to hunting, gathering, agriculture, eating, and art—this is not so much a genus as a superordinate class within which one can distinguish at least doing and making. With Aristotle, such a distinction will be accepted on the basis of "other discourses,"[3] and the specific difference for engineering activity will be sought not within human activity as such but within human productive activity, that is, making.

ENGINEERING AND TECHNOLOGY

It is important at the outset to acknowledge a distinction between engineering and technology. Linguistically, the word *engineer*ing (like *mak*ing or *know*ing) is a gerund, formed by tacking an *-ing* suffix onto a nonfinite verb, and can take objects and adverbial modifiers, thus indicating a kind of activity, whereas with *tech*nology (like *bi*ology or *epistem*ology), the *-ology* creates a substantive that can not take an object, is modified adjectivally, and indicates systematic study or knowledge of that to which it is attached. Solely on the basis of linguistic structure, then, *engineering* connotes activity, whereas *technology* implies knowledge. Attempts to draw inferences about engineering from discussions of technology must therefore be approached with some caution.[4]

In popular usage, however, the English *technology* is seldom defined simply as knowledge and can readily refer to artifacts and activities as well. A discussion of the "technology of travel," for instance, might well mention walking, roads, cars, automobile assembly lines, traffic laws, airplanes, and the aerodynamics of flight. Among different professional groups, the extension of such references can be narrow or broad. In the narrow sense—a sense common among engineers—technology refers to the hands-on or blue-collar work of technicians, even the hardware, that both supports and is subservient to white-collar engineering. In the broad sense—one common among social scientists—technology includes virtually all human making and using of artifacts, even engineering. According to the *International Encyclopedia of the Social Sciences*, for instance,

> Technology in its broad meaning connotes the practical arts. These arts range from hunting, fishing, gathering, agriculture, animal husbandry, and mining through manufacturing, construction, transportation, provision of food, power, heat, light, etc., to means of communication, medicine, and military technology. Technologies are bodies of skills, knowledge, and procedures for making, using and doing useful things. They are techniques, means for accomplishing recognized purposes.[5]

Social scientists treat *technology* the same way they do *culture,* giving it a broad-spectrum, "value-neutral" interpretation.

Even when engineers use the term technology in a broad sense they never get quite so general. The *McGraw-Hill Encyclopedia of Science and Technology,* for instance, defines technology as "systematic knowledge and action, usually of industrial processes but applicable to any recurrent activity" and "closely related to science and to engineering."[6] For the engineer "technology" in a broad sense is scientific technology and does not include craft or art. Only when *technology* is used to denote primarily those forms of making and using artifacts that have been influenced by modern science is it plausible to speak of technology as applied science and engineering as "the technology *par excellence.*"[7]

WHERE ENGINEERING COMES FROM

The term *engineer* (Latin *ingeniator*) was first used in the Middle Ages to designate builders and sometimes operators of battering rams, catapults, and other "engines of war" (or *ingenia*). From this reference to an actor comes the conjugated verb "to engineer" and then the gerund "engineering."[8] Attenuation of the association between engineer and soldier is quite recent. In *Troilus and Cressida,* for instance, Thersites refers to Achilles as "a rare enginer";[9] and in *Paradise Lost* Milton describes "the foe / Approaching, gross and huge, in hollow cube / Training his devilish enginery [= artillery]."[10]

The first professional organization of engineers took place in Milton's time, with the formation of the French military *corps du génie* in 1672. Samuel Johnson's *Dictionary of the English Language* (1755) defines *engineer* as "one who directs the artillery of an army," and Noah Webster's *American Dictionary of the English Language* (1828) describes him as "a person skilled in mathematics and mechanics, who forms plans of works for offense or defense, and marks out the ground for fortifications." Indeed, the first engineering handbooks of the eighteenth century were for bombardiers, and the first schools to grant formal engineering degrees were associated with the military—for example, the Ecole des Ponts et Chaussées was founded in 1747 as an offshoot of the *corps du génie,* the Ecole Polytechnique was founded by Napoleon in 1794 under the direction of the Ministry of War, and the U.S. Military Academy at West Point was founded in 1802 as the first engineering school in the United States. Contemporary distinctions between white-collar engineer and blue-collar technician took a different form among khaki-collar soldiers.

There is, however, a shift in emphasis that takes place between Johnson and Webster; the latter begins to identify the engineer as less the operator (as with those "engineers" who operate already established steam engines and heating plants) and more the one who "forms plans" or thinks things out—albeit with regard to military fortifications. This picks up on a supplementary connotation from the Latin, one that enters English by way of the French. *Ingenerare,* to generate, is what nature does in bringing things into being. Because natural entities exhibit cohesion within themselves (when living, they are sources of motion or, metaphorically, they "work") and coherence with their environment (they "fit"), an artifact that exhibits such properties can be described as *ingeniosus.* Thus the Old French *engignier* is one who contrives or schemes, with intellectual more than just physical force, to make things work or fit—with an implication, perhaps, that they might not otherwise do so. Because of the vaguely impious implications of competing with nature, the fifteenth-century English "yngynore" who plots and lays snares, even though he may well work with his mind and not with his hands, has certain unsavory connotations.[11]

Attempts to equate engineering with traditional architecture— from the Latin *architectus,* transliterating the Greek *architekton,* the prefix *archi-* (primary or master) plus *tekton* (carpenter or builder)—should be viewed with some skepticism, if for no other reason than because of obvious contemporary tensions between the two professions and traditional architectural concerns with urban planning and aesthetic meaning (see Vitruvius, first century B.C.E., *De architectura*).

John Smeaton (1724–92), British architect of the Eddystone Lighthouse and other "public works," was the first person to call himself a "civil engineer"—a term that has come to refer to one who designs roads, bridges, water-supply and sanitation systems, railroads, and so on, that is, publicly funded and utilized projects conceived more from the point of view of utility and efficiency than public space or symbolic expression. (In peacetime armies are also commonly put to building roads, bridges, water-supply and sanitation systems.) In France, the term *ingénieur civil,* which appears in the early 1800s, continues to denote any engineer not employed by the state—whether in mechanics, chemistry, electronics, or other fields. Both terminologies suggest that perhaps no other profession—certainly not those of theology, law, or medicine—has been so well integrated into the state, indicating perhaps a peculiar ability to be co-opted and utilized by independent forces or institutions. The large number of engineers graduated by Japan and the Soviet Union can also be interpreted as support for this point.

ENGINEERING DEFINITIONS OF ENGINEERING AND TECHNOLOGY

As practiced, engineering has come to be defined, in the words of *Webster's New International Dictionary* (1959) and the *McGraw-Hill Dictionary of Scientific and Technical Terms* (1st edition, 1974) as "the science by which properties of matter and the sources of energy [power (McGraw-Hill)] in nature are made useful to man in structures, machines, and products." Ralph J. Smith, an authoritative engineering educator commenting on his own version of this definition—"engineering is the art of applying science to the optimum conversion of natural resources to the benefit of man"[12]— proceeds to conclude that "the conception and design of a structure, device, or system to meet specified conditions in an optimum manner is engineering."[13] Furthermore, "it is the desire for efficiency and economy that differentiates ceramic engineering from the work of the potter, textile engineering from weaving, and agricultural engineering from farming."[14] "In a broad sense," Smith writes later, "the essence of engineering is design, planning in the mind a device or process or system that will effectively solve a problem or meet a need."[15] (The terms *efficiency, effectiveness,* and *economy* will be discussed later.)

Engineering as knowledge is thus identified with the systematic knowledge of how to design useful artifacts or processes—a discipline that, as the standard engineering educational curriculum illustrates, includes some pure science and mathematics, the so-called applied or engineering sciences (e.g., strength of materials, thermodynamics, electronics), and is actualized by some social need.[16] But while engineering involves a relationship to these other elements, still it is designing that constitutes the core of engineering, because it is designing that orders or establishes the unique engineering framework to relate the other elements.

Technology and its cognates are regularly reserved by engineers for more direct involvement with the activity of material construction or the manipulation of artifacts themselves. Actually, engineers tend to take the two cognate chains *technics-technical-technician* (abstract noun, adjective, practitioner) and *technology-technological-technologist* and conflate them to form the grammatical hybrid *technology-technical-technician*. This explains how *technical* and *technician* can be in greater currency when referring to specific practices of making or manipulating in ways that require an adjective or name for the practitioner, while aspects of these same activities can be referred to abstractly as *technology*. As illustrative of this latter use, consider Douglas Considine's *Chemical and Process Technology*. According to the preface, this standard handbook "is *not* a compilation of generalities, but rather is packed with detailed information."[17] Similar examples of this use of *technology* to refer not to theory but to empirical data concerning the raw materials of industrial processes, their equipment, and products can be found in any number of other handbooks.

In light of such a definition, the technician, as the one most directly involved with acquiring and using technical information, is commonly less sophisticated than the engineer. The engineering researcher establishes protocols and methods that the technician employs to collect data; the technician likewise carries out the designs formulated by the engineer. It is this understanding that lies behind, for instance, Smith's distinctions between engineer, scientist, technician, and craftsman:

> The engineer is a man of ideas and a man of action. . . . He develops mental skills but seldom has the opportunity to develop manual skills. In concentrating on the application of science he can obtain only a limited knowledge of science itself. . . . The primary objective of the *scientist* is "to know," to discover new facts, develop new theories, and learn new truths about the *natural* world without concern for the practical application of new knowledge. . . . The engineer is concerned with the *man-made* world. He has primary responsibility for designing and planning research programs, development projects, industrial plants, production procedures, construction methods, sales programs, operation and maintenance procedures and structures, machines, circuits, and processes. . . . The *technician* usually specializes in one aspect of engineering, becoming a draftsman, a cost estimator, a time-study specialist, an equipment salesman, a trouble shooter on industrial controls, an inspector on technical apparatus, or an operator of complex test equipment . . . [The] technician occupies a position intermediate between the engineer and the skilled *craftsman*. The craftsman, such as the electrician, machinist, welder, patternmaker, instrument-maker, and modelmaker, uses his hands more than his head, tools more than instruments, and mathematics and science rarely.[18]

In Great Britain, the term *mechanic* continues to share many of the connotations of *technician*.[19] In France, the different kinds of technical education found in the Ecole Polytechnique, the Ecole Nationale Supérieure d'Arts et Métiers, the Instituts Nationale

Supérieure d'Arts et Métiers, the Instituts Nationaux des Sciences Appliquées and the Instituts Universitaires de Technologie also reflect such distinctions.[20] In the United States distinctions between industrial arts (vocational) education—now more commonly referred to as technology education[21]—and some 1960s proposals to compensate for the increasingly theoretical cast of post–World War II engineering with the creation of a new Bachelor of Engineering Technology (BET) degree,[22] all further illustrate these distinctions.

It is some variation on these distinctions that can be found articulated and defended in such philosophical papers as James K. Feibleman's "Pure Science, Applied Science, and Technology: An Attempt at Definitions"[23] and C. David Gruender's "On Distinguishing Science and Technology." For Gruender, the chief distinction between applied science and technology

> is in the scope or generality of the problem assigned. Those of broader scope we are inclined to think of as problems of "applied" science [= engineering?]; those that are closer to being specific and particular we think of as "technology."[24]

Thus, just as the adjective *technical* connotes a limited or restricted viewpoint, so the engineering technician works from a more limited standpoint than the engineer himself. The technician or technologist might, for instance, know how to perform a test, operate a machine, construct or mass-produce a device (and even be involved in directing others who have a less comprehensive view of some particular operation or construction project), but not necessarily how to conceive, design, or think out such a test or artifact. Consider, for example, the terms "lab technician," "medical technician" and "medical technologist," "drafting technician," and so on. Each designates a person proficient at performing an operation or construction, but not at fully organizing or understanding the activity with which he or she is involved. The engineer is one with a superior or more inclusive view of a material construction than the carpenter or technical assistant.

TECHNOLOGY AS ACTIVITY

Insofar as engineering is part of technology in some broad sense, this technology certainly includes more than knowledge and artifacts, as is readily indicated by its close association with such words as *industry* and *manufacture, labor* and *work, craftsmanship, jobs, operations,* and so on. Indeed, despite the etymology of the term itself and the quickness with which people think of hardware when "technology" is mentioned, activity is arguably its central reality. Technology as activity is that pivotal moment in which knowledge and volition unite to make and to utilize artifacts; it is likewise the occasion for artifacts to influence the mind and will.

Technology as activity is commonly associated with a diversity of human behaviors, distinctions among which are often less clear than is the case with either artifacts or cognitions. Not to mention other difficulties, technological activities inevitably and without easy demarcation shade from individual or personal into group or institutional forms that call for a second, if not wholly independent analysis. Nevertheless, for present purposes attention can reasonably be restricted to individual activities. Among such personal behavioral engagements are

- crafting
- inventing
- designing
- manufacturing
- working
- maintaining
- operating

A cursory inspection of this loose and overlapping diversity suggests that in the active technological engagement with the world there are two broad themes. The first bears on production, the second on use. The former is, as it were, an initiating *action* that establishes possibilities for the latter, recursive *process*. Crafting, inventing, and designing are all actions in technology as activity; manufacturing, working, maintaining, and operating are processes in technology as activity.

The terms here are not wholly satisfactory or firmly fixed but loose linguistic associations that hint at or prefigure certain distinctions that are subject to more detailed exploration and development.[25] Although there are instances for which a phrase such as "making process" or "the action of using" might be employed, a making process tends to be a making action that uses complex technologies, while a using action is oriented toward making and is spoken of as such precisely to stress this aspect. One readily refers to the actions of crafting, inventing, and designing, and the processes of manufacturing, maintaining, operating. To talk about the "inventing process" strikes the ear as slightly off and evidently points to some special form of inventing, "manufacturing action" even more so.

THE ACTION OF MAKING

Setting aside for now the using activity or process as derivative, attention may conveniently be focused on the action of making. Certainly it seems reasonable to think of some kind of making as central to what engineering is. Even if one were to argue the equally central character of using—especially the process of maintaining—this could not displace but only supplement making.

The philosophy of action has largely ignored the unique character of making action and, as Andrew Harrison has argued, is desiccated as a result. In analyses of the rationality of human action, he points out, "the ideas of designing and constructing (except perhaps in somewhat specialized mathematical and related contexts and senses) figure rarely . . . and the no less interesting notions of building, cobbling and bodging not at all."[26] It is true that the philosophy of art occasionally considers aesthetic making, and that creativity (as one specific feature of some making) has been subject to psychological, poetic, pedagogical, and other analyses. Even here, though, the general approach has been to concentrate either on newness, uniqueness, inventiveness or the specific features of inventive cognition. Seldom has making even in this narrow sense been considered simply as activity. Since creative inventing is but one aspect of making in a broader sense, the present analysis will focus on and attempt to identify types of making as human actions.

THREE TRADITIONAL DISTINCTIONS IN MAKING

Aristotle was perhaps the first to suggest a fundamental distinction between two types of making action, that between cultivating and constructing.[27] Cultivating involves helping nature to produce more perfectly or abundantly things that she could produce of herself, and includes the *technai* or arts of medicine, teaching, and farming. Construction, by contrast, entails a re-forming or molding of nature to produce things not found even in rare instances or under the best of circumstances, as with carpentry.

As Andrew G. Van Melsen restates this distinction: In farming,

although man performs all kinds of preparatory tasks, such as clearing, plowing, and sowing, nature itself has to do the rest. Once his preparatory task is done, man can only sit down and wait. It is the inner growing power of living nature which performs the work. [By contrast,] the craftsman gives natural materials forms which would not naturally arise in them. The technical object is something which is not cultivated but constructed, i.e., its component parts are arranged in an artificial pattern. The fashioning of these parts forces them into forms and functions which are not naturally present in them.[28]

Thus, "in the work of construction there is a far more direct intervention in the natural order than there is in the work of cultivation."[29]

Another version of this distinction might contrast technological actions that are in some sense in harmony with nature and those that are not. The environmental and alternative technology movements can be interpreted as a reviving of this cultivation-construction distinction; intermediate or soft technologies assist or imitate nature by acting in harmony with it (utilizing "renewable resources"), whereas hard or high technologies depend on conditions and processes not or only rarely found in nature (using up "nonrenewable resources"). The problem of environmental pollution, at its deepest level, implicates the latter.

Two other traditional distinctions not to be cofused with this are those between the servile and liberal *technai* and between the useful and the fine arts. The servile-liberal distinction depends on whether an action is primarily manual or mental; the useful as opposed to the fine arts depends on another between utilitarian and aesthetic use. The utilitarian art or act of cooking and the fine art of painting (note the modern term *action painting*) are both servile in the classical (pre–eighteenth century) sense, those of theoretical philosophy and practical rhetoric both liberal. Yet independently of such differentiations by actively engaged media or levels of human activity served, it is possible to maintain one between cultivation and construction. Such a differentiation yields a complex, overlapping matrix of servile arts of cultivating (agriculture but not pedagogy) and constructing (carpentry but not engineering) and liberal arts of cultivating (education) and constructing (engineering) as well as utilitarian arts of cultivating (agriculture and pedagogy) and constructing (carpentry and engineering) and fine arts of cultivating (flower growing) and constructing (industrial design).

FROM CRAFT TO ENGINEERING

Singly and in concert, however, these distinctions fail to specify engineering fully. Certainly engineering is not simply a liberal constructing craft, although it may include some craft skills.[30] Harrison's earlier contrast between mathematical designing and constructing and "cobbling and bodging"—compare the American expressions *patching* and *jerry-rigging* and the French *bricolage*—can thus lead to still another contrast. *Bricolage,* the most widely used of these terms, enters intellectual discourse through Claude Lévi-Strauss's attempt to define a practical correlate of mythopoeic thinking. Mythopoeic thinking is a kind of hodgepodge "science of the concrete"; in a similar manner bricolage is a heteronomous collection of specific skills.

> The "bricoleur" is adept at performing a large number of diverse tasks; but, unlike the engineer, he does not subordinate each of them to the availability of raw materials and tools conceived and procured for the purpose of the project. His universe of instruments is closed and the rules of his game are always to make do with "whatever is at hand," that is to say with a set of tools and materials which is always finite and is also heterogeneous because what it contains bears no relation to the current project, or indeed to any particular project, but is the contingent result of all the occasions there have been to renew or enrich the stock or to maintain it with the remains of previous constructions or destructions.[31]

Given some project, Lévi-Strauss goes on to say, the engineer "questions the universe" about how to achieve it. What resources are available? What principles can be utilized? The jerry-rigger, by contrast, simply picks around in a jumble of odds and ends left over from previous activities. Here is something that might work. Try this. In contemporary form he or she is the weekend, *Popular Mechanics* hobbyist. "The engineer is always trying to make his way out of and go beyond the constraints imposed by a particular state of civilization while the bricoleur by inclination or necessity always remains within them."[32] The engineer makes by means of concepts and analysis, the bricoleur by putzing and suggestive accidental combinations. The *bricoleur* is thus inherently more prone to wind up cultivating, the engineer constructing.

In light of this contrast, craft or artisan making can be seen as intermediate between bricolage and engineering. Bricolage is only slightly more "efficient" than nature and "successfully" creates some forms virtually by accident. The putterer patches the roof many times over before it ceases to leak. Action painter Jackson Pollock (who in rebellion against formalized or "engineered" art adopts many of the practices of the *bricoleur*) throws out more paintings than he keeps.

Crafting begins to work its way out of such extreme attachment to its materials through empirical rules and well-developed skills, but in the service of utility integrated with sensuous forms (good tasting foods, tools and clothes with the right feel, warm shapes) rather than strictly visual images or abstract concepts. As a result, there remains a certain continuity with bricolage. Indeed, craftsmen seem to be attracted by the kinds of sensuous complexity of form that arise from improvisation, as has been visually exhibited in the 1964 Architecture without Architects show at the Museum of Modern Art.

The untutored builders in space and time . . . demonstrate an admirable talent for fitting their buildings into the natural surroundings. Instead of trying to "conquer" nature, as we do, they welcome the vagaries of climate and the challenge of topography. Whereas we find flat, featureless country most to our liking (any flaws in the terrain are easily erased by the application of a bulldozer), [these vernacular architects] are attracted by rugged country. In fact, they do not hesitate to seek out the most complicated configurations in the landscape.[33]

But engineering making virtually abandons concern with specific sensuous form (engineering drawings, for instance, are seldom colored, nor do they specify colors) in favor of methodologies of construction that can meet the needs of clients or users, thus reconceiving even traditional cultivation as a kind of construction (witness agricultural and biomedical engineering, not to mention educational technology).

Because of the way it operates within a framework provided by materials it is natural that, prior to the rise of engineering and its abstract conception of making action, making should have been distinguished into different types primarily on the basis of material, cultural, and ritualized formations. Not only is making as bricolage and as craft oriented toward cultivation of nature, but in themselves these activities become cultures. Ethos does not need to strive by means of ethics to impose itself on technical action because technical action transcends its specific cultural roots only in rare instances.[34]

ASPECTS OF ENGINEERING ACTION

Engineering making can be divided into various fields, according to what the engineer actively engages *with,* but also into functions determined by the role the engineer plays in a production sequence. In common parlance engineering functions range from invention, research, and development; through design, production, and construction; to operation, sales, service, and management. Such distinctions are more significant than material divisions because they are repeated within any engineering field.

Invention is sometimes conceived as distinct from and at others as inclusive of research and development. Applied (as opposed to pure) research involves the using of scientific and mathematical knowledge plus experimentation to synthesize new materials or create new energy-generating or transforming processes. Development entails utilizing these materials, energies, and processes to design and fabricate prototype products that solve particular problems or meet specific needs; "industrial research" is another name for this activity.[35]

Design can be considered an activity of development or an activity in its own right ordered toward construction and production. In some instances it is both.

Production and construction are two kinds of making in a restricted sense; the former makes nonstationary artifacts (consumer goods), the latter, stationary structures (houses, buildings). *Fabrication* includes *making* in both senses.

Operation and management denote using processes, as do testing, service, maintenance, sales, and so on—although testing can also be construed as a factor in development and design. The functions of planning, teaching, and consulting cut across these various distinctions.

The relationship among these functions can be schematized in a flow diagram.

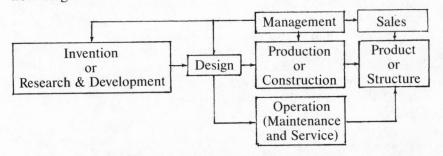

There is, however, no one universally agreed-upon list of such functions; there is simply a spectrum of activities that can be divided and subdivided in numerous ways depending on the kind and degree of analytic detail required. The point is that there are not just electrical engineers, but electrical research engineers (doing applied research on electrical energy or electrical energy-driven devices), electrical design engineers (designing either a specific electrical device for factory production or an electrical system for on-site construction), electrical service engineers (maintaining and servicing some electrical product or system), and so on.[36]

APPROACHES TO INVENTING

As an aspect of engineering action, inventing can be approached by means of selective contrasts.

As *opposed to scientific discovery,* technological inventing refers to the creating of something new rather than the finding of something already there but hidden. Alexander Graham Bell invented the telephone; Isaac Newton discovered the law of gravity. The telephone did not exist prior to Bell's work; gravity existed but was not conceptualized in the form of scientific law prior to Newton. Adopting the realist epistemology characteristic of scientists and engineers themselves, invention causes things to come into existence from ideas, makes world conform to thought; discovery, by deriving ideas from observation, makes thought conform to existence. One difficulty with this view is that scientific ideas (theories) are underdetermined by observations and require some conceptual or imaginative creativity.

As *opposed to conceiving* or even *imagining,* it is the concrete transformation of materials—the making physically real of an imagined transformation—that is the essence of inventing. Sir George Cayle (1773–1857), founder of the science of aerodynamics, had an accurate conception of the airplane, but its invention (dating from the first successful flight at Kitty Hawk in 1903) had to await both the development of a suitable power plant and the Wright brothers' technical skills (of fabricating and operating). Did Leonardo invent the parachute merely by imagining it, or Lenormand by fabricating and testing it? Inventing may begin in some conceptualization, but does not finally take place until an artifact is operationally tested and discovered able to perform its assigned task.

As conscious action originating in the mind and confirmed by worldly engagement, the concept of invention is a distinctly modern notion[37] and is *opposed to slow* or *incremental changes* in techniques. As with scientific discovery, invention can take place over a short period of time in a single individual who introduces historical discontinuity, or it can be prepared through gradual development within a group. The slowed-down or spread-out invention through innumerable minor modifications maintains historical continuity and has sometimes been termed *innovation* (although this term now more commonly specifies the bringing to market of an invention).[38] Other observers emphasize the historicosocial character of inventing,[39] replacing individual inventors with technical communities, national groups, or historical periods as inventors of such artifacts as the astrolabe or compass.

As *opposed to designing,* inventing appears as an action that proceeds by nonrational, unconscious, intuitive, or even accidental means. Designing implies intentionality, planning. Inventing is, as it were, accidental designing—and as such highlights the element of insight and serendipity that plays an important role even in highly systematized design work.

FROM INVENTING TO SYSTEMATIC INVENTING

Modern engineering, as an attempt to settle and systematize the inventive process, has been called the "invention of invention."

The greatest invention of the nineteenth century was the invention of the method of invention. A new method entered into life. In order to understand our epoch, we can neglect all the details of change, such as railways, telegraphs, radios, spinning machines, synthetic dyes. We must concentrate on the method itself; that is the real novelty, which has broken up the foundations of the old civilization.[40]

A key figure in this development was Thomas Edison (1847–1931), who in the 1870s established what he called an "invention factory" that would make "inventions to order." For Edison, inventing is the product of organized purpose. Although by the early 1800s inventing was well recognized and culturally prized, it remained based largely on individual initiative and intuition, and divorced from direct large-scale organization or financial backing. It was Edison who, especially with his massive, methodically directed trial-and-error search for a suitable filament for the incandescent light in conjunction with the systematic development of related elements necessary to its commercial exploitation (vacuum bulbs, parallel circuits, dynamos, voltage regulators, metering devices) first created the industrial research organization tied to capitalist economic structures.[41]

Inventing and systematic inventing (sometimes called the designing of inventions) are sometimes contrasted by saying that an inventor creates the new whereas the engineer plans and discovers the possible. An engineer remains within the familiar and systematic, does not venture into the unknown, only orders or re-orders the known along well-established methodological lines, so that, given a clearly specified problem, two equally competent engineers will discover approximately the same solution.

Friedrich Dessauer, however, has argued that inventing or creating also involves the experience of discovery.[42] Indeed, the word *to invent,* from the Latin *invenire,* means *to come upon, to find,* or *to discover.* Moreover, inventing is capable of exhibiting parallel histories and objective confirmation—for example, when two persons independently invent the same thing (as with Elisha Gray and Alexander Bell, who both invented the telephone, and even applied for patents on the same day). Gilbert Simondon provides a detailed mechanology or descriptive phenomenology of machines, documenting the tendency of inventing to generate certain stable forms, especially insofar as these forms are not obscured by the play of fashion and commercialization.[43] So much is this element of discovery and objectivity present that to explain it Dessauer feels justified in postulating the existence of a transcendental realm of pre-established solutions to technical problems. The natural or external world explains or accounts for the objectivity of science; but since inventing does not bear upon what already is, there must be a transcendent *is-ness* or being to account for its discoveries.

Others, however, while agreeing with Dessauer on the moment of discovery in invention, appeal to less metaphysical explanations. David Pye, for instance, argues quite simply that

Invention is the process of discovering a principle. Design is the process of applying that principle. The inventor discovers a class of system—a generalization—and the designer prescribes a particular embodiment of it to suit the particular result, objects, and source of energy he is concerned with.

The facts which inventors discover are facts about the nature of the world just as much as the fact that gold amalgamates with mercury. Every useful invention is a discovery about the way things and energy can behave. The inventor does not make them behave as they do.[44]

Previous contrasts with inventing throw further light on the movement from simple or accidental to systematic invention. Whereas primitive inventing relies on accident, bricolage, fortuitous insight into possible relationships among elements in the given, invention research develops a calculus of such relationships that can be used to solve well-specified problems. The fact that such a calculus may still rely at crucial moments on a cultivated serendipity (brainstorming sessions) and heuristics only reveals that irreducible essence of invention as creative insight which must so far remain as a circumscribed aspect of systematic invention.

For summary purposes the making action in the initial instance (it will be different with routine making) can be broken down into the following sequence:

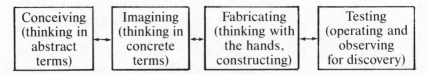

This is a logical, not a historical sequence; the various moments will obviously be existentially interrelated in considerably more complex fashion than can be indicated schematically. Nevertheless, to "invent invention" is to set up an institution that enhances the conditions under which these various moments in the making action can be pursued and can interact. Industrial research and development laboratories or applied research institutions are the result.

ON ENGINEERING DESIGN

The second moment in the above sequence can be identified as the location if not yet the essence of engineering design.

As previously indicated, virtually all general articles on engineering and all introductory engineering textbooks identify designing as the essence of engineering. The design project is typically the capstone of an undergraduate engineering education. The most well-developed field of research on engineering qua engineering—as opposed to research within some branch of engineering—deals with engineering design and especially design methodology.[45] When engineering method is contrasted with the scientific method it is the method of design that is invoked.

But what, exactly, is engineering design? Designing (from the Latin *designare*, "to mark out") specifies some material object in sufficient detail to enable it to be fabricated. It is, as it were, reified intention. Indeed, the very word can be used as synonymous with intention (as in "his design was to make more money"). The problem is that the standard engineering definitions of designing do little more than rephrase the standard definitions of engineering itself. Examples: "Engineering design is the process of applying the various techniques and scientific principles for the purposes of defining a device, a process or a system in sufficient detail to permit its physical realization."[46] Or, engineering design is "an iterative decision-making activity to produce the plans by which resources are converted, preferably optimally, into systems or devices to meet human needs."[47] Or yet again, engineering design "is the intellectual attempt to meet certain demands in the best way possible."[48] Designing may thus be described as the attempt to solve in thought, on the basis of available knowledge, problems of fabrication that will save work (as materials and/or energy), either in the artifact to be produced, the process of production, or both.

Consider, for example, a foundation for some structure. Were a stonemason to construct this foundation on the basis of experience and intuition alone, one of two things is likely to happen: either it will be made too weak, so that the building eventually collapses and has to be rebuilt; or, what is more likely, it will be made too strong, utilizing more stone and concrete and steel than necessary. In either case, more work than needed will have been done. Were an engineer to design the same foundation, there would be an attempt to calculate the weight of the building and other relevant forces, then using the principles of physics plus engineering geology (i.e., geological knowledge interpreted in terms of what kinds of structures various earth formations can support), and a socially specified safety factor, and the foundation would be formulated with neither more nor less than what was required. Although paradoxical, the right construction (like Aristotle's golden mean) is difficult to attain; it takes effort. But when this effort is expended at the right time, in the long run it saves effort. Engineering design is thus an effort (at first sight, of a mental sort) to save effort (of a physical sort).[49]

This mental effort is, however, something distinct from knowing or coming to know in a scientific or theoretical (or even technological) sense, because it terminates not in an interior cognitive act, but in construction—that is, in miniature making. Scientists often experience a tension between their knowledge and what they can express; they make discoveries and then have to push beyond what they feel is their proper sphere in order to write them up. But such tension is not a normal feature of design experience, because the construction of drawings or models (which also serve communication) is intimately bound up with the design process.

Joseph Edward Shigley, in a standard book, *Theory of Machines*, argues that "the use of the drawing board in kinematics instruction is very desirable and usually necessary" because "the most direct method of attacking a kinematic or dynamics problem is the graphical one."[50] If a design project is the capstone of an undergraduate engineering education, it ultimately rests on the foundation of engineering graphics, required of all freshmen. The importance of the physical act of drawing is also indicated by another author's contrast between sensorimotor skill, phantasmal capacity, and conceptual capacity. "The *phantasma*, or sensory representation at whatever level of complexity . . . is what we are concerned with."[51]

Engineering drawings, with their unique language and system for abstraction and representation, are not just means for communicating results arrived at by interior activity; they are part of the process and the means by which the results themselves are reached. At the same time, drawing is only one way of performing the more general engineering action of modeling.

> One source of confusion in thinking about design is the tendency to identify design with one of its languages, drawing. . . . Design, like musical composition, is done essentially in the mind, and the making of drawings or writing of notes is a recording process. The designer, however, uses drawing for self-communication just as everyone uses words for thinking. This use of drawing as an extension of the mind, a sort of external (and reliable) memory can be a very important part of the design process. Drawing should be taught not primarily to give the student facility in the use of tools—pencil, triangle, tee square, and most important, the eraser—but to give him practice in pictorial extension of the mind. It is not to be expected that all students are equally endowed with the ability to think pictorially any more than to think mathematically. Somehow educators tend to look upon mathematical ability as a more desirable quality than the ability to think in terms of spatial relations. Before dismissing the latter as something of lesser merit, it may be well to reflect that one of the greatest engineers of all time, Leonardo da Vinci, was essentially a draftsman, not a mathematician.
>
> Pictorial language is especially well adapted to expressing particular physical form and physical space relationships. Functional relationships are often better expressed by a symbolic language. Such languages have particular facility in expressing generalizations without specifying detail. The chemical engineer uses the flow sheet, the electrical engineer the circuit diagram, and all kinds of engineers use the block diagram as important tools in the conceptual process.

The designer often uses the symbolic languages of mathematics but usually in connection with the analysis of a design rather than directly in the conceptual process.[52]

Although it starts out by stating a position ostensibly at odds with the idea of design as essentially involved with drawing or miniature construction, by arguing that drawing is merely one form of modeling or picturing, this passage actually supports and generalizes the present argument.[53]

The point at issue is confirmed by yet another analysis of "technology as knowledge." After noting how for artists thinking means something different than for philosophers, and how "technologists display a plastic, geometrical, and to some extent nonverbal mode of thought that has more in common with that of artists than that of philosophers,"[54] Edwin T. Layton, Jr., describes designing in the following terms:

> The first stages of design involve a conception in a person's mind which, by degrees, is translated into a detailed plan or design. But it is only in the last stages, in drafting the blueprints, that design can be reduced to technique. And it is still later that design is manifested in tools and things made. Design involves a structure or pattern, a particular combination of details or component parts, and it is precisely the gestalt or pattern that is of the essence for the designer.
>
> We may view technology as a spectrum, with ideas at one end and techniques and things at the other, with design as a middle term. Technological ideas must be translated into designs and tools to produce things.[55]

Layton's only mistake here is to call this designing activity primarily a kind of knowledge and to fail to notice that modeling in one form or another is something that goes on, not just at the stage of making blueprints, but at virtually all stages of engineering action. Indeed, this activity can be described as the creation of a series of designs, first quite general (freehand sketches, perhaps simply a block diagram analysis of the problem), but progressively detailed and specific (working drawings), until it terminates in the actual construction (a process overseen by engineers in which the carpenters and other "technicians" act to some extent like "living pens and pencils" scaling up a drawing or design one last time). Generous doses of trial-and-error learning, intuition, and practical skill are regularly involved.

With regard to the last suggestion, although the actual execution of a plan is not designing, except insofar as the plan may continue to be modified in order to meet originally unanticipated situations, such continuing development through final execution is in fact the norm; on large construction projects a draftsman will be continuously at work revising drawings in the light of exigencies and changed circumstances, thus seeking to anticipate their further consequences. Drawing is a kind of testing or interrelating of various factors[56] by miniature building. It is not thinking in the sense of conceptualizing or the relating of concepts; it is thinking as picturing or imagining, and the relating of specific materials and energies. The designer solves problems of relating parts the way an artist does, by seeing them in practice. It is activity, only on a reduced scale, made as free of physical labor as possible, but nevertheless not entirely free. It is still an effort (of a miniature physical sort) to save effort (of a gross physical sort).[57] This particular miniaturization of construction is, however, intimately related to special kinds of knowledge, especially the engineering sciences.

EFFICIENCY, EFFECTIVENESS, ECONOMY

Engineering is a systematic effort to save effort. But what, more precisely, is this system? Perhaps it can be further explicated as the pursuit of efficiency? As a guiding principle of engineering design, engineers themselves repeatedly refer to the ideal of efficiency. Jean-François Lyotard refers to all technology (including, presumably, engineering) as "a game pertaining not to the true, the just, or the beautiful, etc., but to efficiency."[58] But what is efficiency? As Stanley Carpenter and others have well pointed out, efficiency is a context dependent variable.[59] Context dependency, however, need not deprive a concept of all formal characteristics; even variables can have boundaries. Legal justice, for instance, is equally context dependent, but nevertheless denotes a special way of looking at behavior, that is in terms of its conformity with a set of rules articulated and enforced by the state. Engineering likewise can be said to have its own special perspective on the activity of designing—in terms of efficiency.

The term *efficiency* has its roots in the Latin *efficere* (to produce, effect, or make). The adjectival derivative, *efficiens,* modifying *causa,* indicates one of the Aristotelian four "causes of motion."[60] In contrast with formal, material, or final causation, the efficient cause is the "principle of change" that unites the other three. In English, *efficiency* traditionally indicated the operative agency or power of something or someone to get something done, to produce results. In Christian theology, God as creator *ex nihilo* is described as the supremely efficient or "most effective" cause. In military parlance of the late 1800s a soldier who could do his job was "efficient" or, it would be said today, "effective." In all such uses there is no sense of efficiency in the technical or engineering sense.

The first use of the word *efficiency* in what has come to be called the technical sense occurs, appropriately enough, in mechanical engineering during the second half of the nineteenth century,[61] although its roots are no doubt in what Pascal refers to as "l'esprit géométrique."[62] In the technical sense, efficiency is defined as a ratio of outputs to inputs. It is difficult for us to recognize how unusual this perspective is.

The idea of looking at or judging any object or process in terms of a relationship between outputs and inputs is not, for instance, operative in our primordial engagements with quotidian practices such as speaking, dressing, eating, and so on, nor in any premodern system of morality. Common speech and traditional rhetoric are prolix, wasteful, inefficient in their effusiveness; we say the same thing over and over again in different ways. Only modern rhetoric is pared down to essentials, aspires to say no more than is necessary. We normally eat what tastes good or what is appropriate for an occasion, not what gives us the right amount of nourishment for the least expense of money, time, or effort. For Plato one should strive to imitate an ideal; for Aristotle the goal is a more immanent virtue or perfection of operation. Natural law theory likewise stresses acting in harmony with some larger order.

From its origins, mathematization of the practical world—in contrast with mathematical theory—has been recognized to exhibit a paradoxical character. On the one hand, it introduces into human affairs a new level of accuracy and rigor; on the other, it tips human experience and attention away from the most important things. Those sciences that are crucial to education such as music, medicine, agriculture, and leadership, necessarily rely, suggests Plato, on cultivated intuition. Shipbuilding, house construction, and woodworking, by contrast, employ lathes, compasses, chalklines, and all sorts of ingenious instruments.[63] One especially powerful and influential form of mathematicized practice is what is today termed input-output analysis.[64]

Thinking in terms of an input-output relationship is first explicitly presented in the pseudo-Aristotelian treatise, "Mechanical Problems,"[65] and subsequent Archimedean analyses of simple machines (i.e., tools). Indeed, the very concept of an *instrumentum* as an object to be judged in terms of its "use," independent of user and context—a notion related perhaps even to the idea of the sacraments as "instruments of salvation," the efficacy of which is independent of the moral character of the priest administering them—implicitly entails input-output considerations. But the idea only begins to take on clearly definable form with early modern mechanics,[66] the advent of double-entry bookkeeping, and the formulation of theories of political economy. Certainly the notion plays no role in Aristotle's *Oeconomica* or other premodern economics texts. The amplification and application of efficiency in management and as a technical concept in economics was also undertaken by engineers, one turned manager (Frederick W. Taylor) another turned social scientist (Vilfredo Pareto),[67] and its popularization as a social ideal was promoted by engineers turned social-political activists.[68] The moral theory of utilitarianism can also be interpreted as incorporating and able to be integrated into a general theory of action based on efficiency.[69] Contemporary arguments that benefit (outputs) have not only costs (inputs) but risks (as either outputs or inputs), and the observation that progress (a benefit) has costs, are sophisticated extensions of efficiency and input-output analysis.

The philosopher who has expended the most effort to elucidate the character of efficiency as a technical ideal in engineering is Henryk Skolimowski.[70] According to Skolimowski, progress in science is indicated by better theories (increases in knowledge), progress in technology by better artifacts or the processes for making artifacts (increases in effectiveness or efficiency). Skolimowski then tries to indicate the specific forms effectiveness takes in different branches of engineering. In surveying it is accuracy of measurement, in civil engineering, durability of structures, and so on.

But Skolimowski's analysis is deficient in two respects. It does not properly distinguish effectiveness and efficiency. The two terms are not interchangeable, as they are often treated, nor is efficiency "a measure of effectiveness." A less efficient but more powerful bomb could easily be more effective than a more efficient but less powerful one. Either could be more or less economical—that is, efficient in monetary terms—to develop or manufacture.

Skolimowski further fails to indicate the more subtle and sophisticated forms of efficiency able to be embodied in engineering design methodologies. For this, one can turn to the work of Herbert Simon, who describes the science of design as including the methods of optimization and "satisficing" (also called "bounded rationality"). A logic of optimization, for instance, distinguishes between what are called command variables (means or inputs), fixed parameters (laws or rules), and constraints (ends or outputs). "The optimization problem is to find an admissible set of values of the command variables [e.g., kinds and quantities of food], compatible with the constraints [nutritional requirements], that maximize the utility function" for some given environment or situation.[71]

This need not always produce an "absolute efficiency" or some "one best way."[72] Given the complexities of most real-world problems, it is sufficient if there exists a design method that provides a way for choosing between a number of alternatives or for doing what Simon calls "satisficing," i.e., achieving a satisfactory if not perfect solution to a problem—one that is more efficient than others within the bounds of those with which it can be compared. This process can entail cost-benefit analyses even of the process of design itself.[73]

The creation of this "science of the artificial" has transformed engineering design from an "intellectually soft, intuitive, informal, and cookbooky"[74] activity to one characterized by "optimizing algorithms, search procedures, and special-purpose programs for designing motors, balancing assembly lines, selecting investment portfolios, locating warehouses, designing highways, and so forth."[75] The counterargument of Donald Schon, that engineering design retains the character of "a reflective conversation with the materials" of a situation,[76] overlooks both the distinctive character of what this conversation is about and its special structure. In a like manner, Billy Vaughn Koen's stress on the role of heuristics in engineering design fails to acknowledge the systematic character of engineering heuristics.[77] Koen is no doubt correct that there is not always some universal solution to a problem, but this is more because that problem does not have a truly unique form than because the engineering method has room for multiple solutions. Were Koen's argument pressed, it would return engineering to bricolage.

Although it is true that what can be counted as inputs and outputs can be virtually unlimited and is often socially determined, within some specified input-output parameters, engineering design searches for ways to minimize the input-output difference[78]—by means of miniature construction. In German the very word for engineering design, *Konstruktiontätigkeit* (literally, "construction activity"), confirms this point. This miniature construction proceeds most often by means of visual representation, but also by modeling and by mathematical analysis of the resulting drawing or model. It is the special visual, schematic representation[79] of this input-output conceptualization that definitively characterizes engineering design and allows it to proceed under the ideal of efficiency—that is, making choices between alternatives on the basis of comparisons of input-output relationships—thus distinguishing it from other types of designing.

DESIGNING IN ENGINEERING AND IN ART

In contemporary theory of engineering design, the structure of this action is thought to be describable as a method, differing from yet analogous to the scientific method of knowing—that is, as a method of practical action. As such it has been argued to underlie all practical activity not only in engineering but in business, education, law, politics, art—if not simply all human action. The method is one; the only differences are in goals pursued—and, perhaps, in materials employed. One designs business ventures to make money, educational curricula to impart information and knowledge, laws to alter public behavior and be enforceable, political campaigns to win votes. At the same time, it is primarily within the engineering field that the methodology of design has been most seriously investigated.[80]

Without going into detail regarding the discussion of the exact character of this method itself or its relationship to the method of science, one can nevertheless distinguish engineering from artistic design on the basis of ideals or ends in view. In contrast to the engineering-design ideal of efficiency, there is the artistic design ideal of beauty. Beauty is not so much a question of materials and energy as one of form.[81] About this the whole subject of aesthetics has more to say, whereas it is ethics or politics that would incorporate a philosophical evaluation of efficiency.

Yet the difference between these two types of design does not remain at the level of ideals; it penetrates to the design activity itself. This is apparent first in the way efficiency refers to a process, is a criterion for choosing between processes or products conceived as functioning units, whereas beauty is in the primary instance a form of stability. Does a potter aim at efficiency in creating a beautiful pot? No, the aim is a good work, one of proportion and harmony; efficiency in production, while not to be wholly ignored, is a distinctly secondary consideration. For the engineer, however, it is beauty that is of secondary importance; while not to be ignored, beauty is judged, in industrial design, in terms of its contribution to function or efficiency, perhaps even to marketability or sales.[82] Even more clearly: the ends of artistic design must be formal whereas the final causes of engineering actions are readily susceptible to verbal articulation in terms of human needs, wants, desires, and so on.

A further observation: Engineering design limits itself to material reality (metaphysically, matter and energy are both matter as contrasted with form). This limitation is to be grasped or approached, however, by means of a mathematical calculus of forces derived from classical physics (Galileo and Newton) and its specific mathematical abstraction. The picturing or imagining that goes on in engineering design is done, as it were, through the grid of this physics—the grid itself being articulated as the engineering science of mechanics. This viewing of matter and energy through the grid of classical physics gives to engineering design a rational character not to be found in art. Engineering images, unlike other images, are subject to mathematical analysis and judgment; this is their unique character and one that sometimes leads people to confuse them with thinking in a deeper sense.

Art also is concerned with imagining, but its images cannot be rationally analyzed; they are not subject to any well-developed calculus. Thus art, in contrast to engineering, appears as both more intuitive and more dependent on the senses. Although the artist, too, is concerned to design artifacts, he or she necessarily does so in drawings and models that remain close in being to the final product. Compare, for instance, a Rembrandt sketch for a painting with an engineering drawing of a building (which is not to deny that architectural drawings often do exhibit a kind of aesthetic character).[83]

THE NON-NEUTRALITY AND NON-AUTONOMY OF ENGINEERING

Responding, in the first instance, to popular cliché and, in the second, to critics such as Jacques Ellul, two negative theses have developed regarding the basic character of modern technology: Technology (notwithstanding the contentions of Jacques Ellul, Langdon Winner, and others) is not autonomous. The positive form of both theses is this: Technology is a value-driven social process.[84] The assumption is also commonly made that what is said about technology *ipso facto* applies to engineering.

The problem with the idea that engineering is a value-driven social process is that either it is too simply true or it simply redefines a basic question. All human activities are value driven in the sense of being choice behaviors. In this sense "value" refers to any motivation, taste, preference, or decision one way or another on any basis whatsoever. On the argument of Emmanuel Mesthene,[85] technology increases possibilities and thus intensifies the process of making choices or valuing. Certainly engineers have to make thousands of choices in the course of designing the simplest device. But all this tells us is that engineering is a human activity; it does not yet contribute to an understanding of what kind of activity engineering is.

Prescinding from such a vacuous interpretation, the crucial question becomes, What kinds of values drive engineering in some unique way and/or what unique values drive engineering in some quite common way? Here one can distinguish two basic views: that the values are internal to engineering or that they are external. A strong restatement of the externalist position (which goes back at least to Karl Marx) has been made by Steven Goldman. According to Goldman,

> What is typically the case is that a knowledge base in some branch of engineering or science is capable of evolving in a number of directions and/or is capable of being the basis of a wide range of concretizations as products or production techniques. The determination of which direction development of the knowledge base will take, or which of a range of possible products will be realized, is made on the basis of economic, institutional, political, social, and sometimes personal value judgments.[86]

Goldman nevertheless admits that "internally and in its relationship to society, technological action has a logic of its own" and suggests that "it is precisely in explicating the logic of this process, within which the determinants of action derive from non-unique value judgments rather than from either the products of technology or its knowledge base, that an understanding of technology lies."[87]

The thesis here is that the internal value constitutive of and operating in engineering is the ideal of efficiency, and that the pursuit of efficiency thus determines the ways in which external economic, institutional, political, social, and sometimes personal value judgments can influence engineering. Crudely oversimplifying, for any external value to influence engineering it must be stated in the form of an input or an output. Thus, although modern engineering is indeed subject to diverse social transformations, all of them are necessarily in tension with such nontechnical ideals as living in harmony with a traditional way of doing things, the imitation of spiritual or transcendent patterns, the development of virtues, and deontological claims regarding the limits of human action.

The situation is analogous to that created by the introduction into a social order of the ideal of legality. Although law is subject to transformation by diverse external values, nevertheless all societies governed by law share certain common features and differ in important respects from social orders based on custom and mores. Customs that cannot be stated in legal form tend to lose their binding force. In a comparable manner, input-output or efficiency-oriented thinking destabilizes traditional societies and tends to introduce a new way of being that exhibits certain common features no matter what particular external values influence it. Research and development laboratories, for instance, certainly exhibit a common internal culture throughout the world; not unlike Holiday Inns or McDonald's Restaurants, they are the same wherever they are found. Engineers and scientists regularly express their transnational and transcultural affinities,[88] and in fact they find it very easy to move from one country to another. Indeed, the very ease with which engineering can often be inserted into some cultures and co-opted by various power interests—and the difficulties of doing this in other cases—likewise suggests a distinctive and influential internal character.

Any metaphysics of engineering, like metaphysics generally, seeks to account for both sameness and difference. In the case of engineering, sameness refers to the ways in which engineering is one throughout its various manifestations, difference to the way these manifestations can become means for diverse external ends. In the case of engineering, it is further necessary to explain how and to what extent an ostensible means truly amenable to various external values can also on occasion become an end in itself.

ETHICAL IMPLICATIONS

Metaphysical observations have implications for ethical analysis. To begin with, the idea of engineering as action is the ground of engineering ethics. Only if engineering is activity (not primarily knowledge), is there a prima facie case for ethics in engineering. It is ironic that while philosophers and historians of science are learning to think of science not so much as knowledge but as action,[89] many historians and philosophers of technology have been trying to think of technology less as action and more as knowledge.[90]

Second, that design is the core of engineering activity points toward a different emphasis in professional engineering ethics than has been characteristic of leading contemporary developments. To date most case studies in engineering ethics focus not on designing itself but on pre- and postdesign activities, and on the need to let engineers be engineers. The standard defense of the whistle-blower is basically a defense of the engineer and his own inherent standards. It assumes that the design activity is sound if it is just left alone, if it is not distorted by commercial or political interests. Like lawyers and physicians, engineers are encouraged to pursue and defend a kind of professional autonomy without questioning the inner character of such an "autonomous" profession.

Insofar as there have been discussions of the ethics of design, in contrast with the ethics of the applications of design, arguments have called for an expansion of input-output considerations.[91] Christian engineer Lambert Van Poolen, for instance, has argued for a "technology design philosophy" that would "embrace all aspects of reality," and for replacing the concept of technical efficiency with one of holistic sufficiency.

> The concept of "sufficiency" in design is meant to supplant the narrow concept of technical, financial, and marketing "efficiency" so often used as guides for the development of design specifications. [T]hese categories of efficiency criteria . . . are not adequate or sufficient for proper specification of a designed tool or product operating in concrete, everyday reality. The call is for an expansive consideration of the entire life-cycle of technological objects with concomitant proper development of a broad range of specifications in light of design norms.[92]

It is not clear, however, that an input-output calculus, no matter how sophisticated, can do justice to many notions of goodness, rightness, beauty, and truth beyond those adaptable to utilitarian assessments. In fact, there is little if any empirical evidence that in cases amenable to both nonengineering and engineering design strategies the latter actually improves the design. Just as the professions of law and medicine have come under some scrutiny as law and as medicine—that is, questioned about their own inner character, with proposals for the external limitation (not just redirection) of some of their activities—so too may engineering be legitimately subject to such restrictions, especially in the research and development setting. Certainly this has been raised as a possibility with regard to modern engineering research in fields such as nuclear development and biotechnology. The full articulation and defense of such a suggestion is, however, beyond the scope of the present preliminary analysis.

NOTES

1. Cicero, *Tusculan Disputations* 5.4. 10–11; Plato *Apology of Socrates* 22d–e; and Xenophon *Memorabilia* 3.10.

2. For a treatment of technology as object, as knowledge, and as volition, see Carl Mitcham, "Types of Technology," in *Research in Philosophy and Technology* (Greenwich, Conn.: JAI Press, 1978), 1:229–94. Although not in agreement with the thesis of that paper, the present study nevertheless incorporates substantial material from section 2, on technology as process (activity), appropriately revised, and presumes the extensive references provided there.

3. Aristotle, *Nichomachean Ethics* 6.3. 1140a4. But see, for example, Hannah Arendt, *The Human Condition* (Chicago: University of Chicago Press, 1958); Nicholas Lobkowicz, *Theory and Practice: History of a Concept from Aristotle to Marx* (Notre Dame, Ind.,: University of Notre Dame Press, 1967); and Yves R. Simon, *Work, Society, and Culture,* ed. Vukan Kuic (New York: Fordham University Press, 1971).

4. It is remarkable that the classic nonengineering philosophers of technology such as José Ortega y Gasset (1883–1955) and Martin Heidegger (1889–1976) seldom mention engineering as such. With the classic engineering philosophers of technology such as Ernst Kapp (1808–96) and Friedrich Dessauer (1881–1963) the situation is, of course, quite different.

5. Robert S. Merrill, "The Study of Technology," *International Encyclopedia of the Social Sciences* (New York: Macmillan, 1968), 15:576–77.

6. *McGraw-Hill Encyclopedia of Science and Technology,* 5th ed. s. v. "Technology." This article has remained unchanged from the first edition.

7. Mario Bunge, *Treatise on Basic Philosophy,* vol. 7, *Epistemology and Methodology* 3: *Philosophy of Science and Technology* (Dordrecht: Reidel, 1985), p. 241.

8. Notice that this is just the opposite of how the term *scientist* (the name for one who does science) is derived from *science* (the thing done), which—despite a strong English tendency to turn nouns into verbs—cannot be conjugated. *Scientist,* incidentally, originates much later than *engineer,* being documented first in William Whewell's *Philosophy of the Inductive Sciences* (1840).

9. *Troilus and Cressida,* act 2, sc. 3, line 8. In Shakespeare see also "the engineer / Hoist with his own petard" (*Hamlet,* act 3, sc. 4, line 206). According to the *Oxford English Dictionary,* 2d ed. (1989) similar usage can be documented as early as 1325.

10. John Milton, *Paradise Lost,* bk. 6, line 553.

11. Richard Sibbes's commentary in Charles Haddon Spurgeon's *The Treasury of Psalms* (1635) Psalm 9, line 15, refers to "that great engineer, Satan." And of course Milton's Lucifer is the engineer of Pandemonium, a city built to rival God.

12. Ralph J. Smith, Blaine R. Butler, and William K. LeBold, *Engineering as a Career,* 4th ed. (New York: McGraw-Hill, 1983), p. 9. For a sample of definitions running from that of the British architect and civil engineer Thomas Tredgold

(1788–1829) to the Accreditation Board for Engineering and Technology (1982), merge Smith, 3d ed. (1969), pp. 8–9, with Smith, 4th ed. (1983), p. 8.

Note: The first three editions (1956, 1962, 1969) were all authored by Smith alone. Although he calls the 4th edition "primarily the work of coauthors," a brief comparison reveals nothing new in the quoted passages; hence the accreditation to Smith alone in the text.

13. Smith, *Engineering as a Career,* 4th ed., p. 10.

14. Ibid., p. 12.

15. Ibid., p. 160. Thomas T. Woodson's oft-cited *Introduction to Engineering Design* (New York: McGraw-Hill, 1966), while noting that not all engineering functions involve design in the strict sense, also identifies *engineering design* as "the essential activity of professional engineering" (p. 8). Variations on this thesis can be found in virtually all general engineering texts.

16. The category of social need or human benefit is more complex than might initially appear. Although an extended discussion is not possible here, it is nevertheless relevant to note two distinctions. Social need, which motivates society to dedicate resources to engineering, is not necessarily the same as what motivates individual engineers (cf. the love of tinkering with machines, etc.). Furthermore, the presence, even the recognition of a social need is not in itself sufficient to motivate society to promote engineering; it is also necessary that engineering be seen as a legitimate means—one not in conflict with other societal responsibilities—and perhaps even the world as composed of "resources."

17. Douglas M. Considine, ed., *Chemical and Process Technology* (New York: McGraw-Hill, 1974), p. xxii. As specified on the previous page, this detailed information concerns "the traditional spheres of interest in industrial chemistry and chemical technology as reflected by the petroleum, petrochemical, chemical, paper, textile, and other long-established process industries" as well as "more recent applications of an advancing and broadening chemical technology, including, as examples, the materials and processes now required by the electronics, optics, and aerospace industries." A pie diagram on p. xxviii gives a conceptual overview of the percentages of different kinds of information included: roughly one third is devoted to data on raw materials, one third to equipment specifications, and one third to consumer products descriptions.

18. Smith, *Engineering as a Career,* 3d ed., pp. 210–11. The earlier edition is cited because in the 4th, in order to avoid the sexist implications of referring to the engineer as a "man" and "him" or "his," the terms are always either "engineers" or "engineering" and "their," and the prose has become too contorted for easy quotation. Cf. also Philip Sporn, *Foundations of Engineering* (Oxford: Pergamon Press, 1964), pp. 18–19, which further distinguishes between "technologist" and "technician."

19. "The man who fixes the gas cooker or wires the house for electricity is not an engineer. He is a mechanic or an electrician." (Wiliam T. O'Dea, *The Meaning of Engineering* [London: Scientific Book Club, 1961], p. 11.) See also the "mechanics institutes."

20. See John Hubbel Weiss, *The Making of Technological Man: The Social Origins of French Engineering Education* (Cambridge: MIT Press, 1982); and Charles R. Day, *Education for the Industrial World: The Ecoles d'Art et Métiers and the Rise of French Industrial Engineering* (Cambridge: MIT Press, 1987).

21. For this perspective, see Paul W. DeVore, *Technology: An Introduction* (Worcester: Davis Publications, 1980).

22. John Dustin Kemper, *Engineers and Their Profession,* 3d ed. (New York: Holt, Rinehart and Winston, 1982), p. 277. For further documentation see James J. Duderstadt, Glenn F. Knoll, and George S. Springer, *Principles of Engineering* (New York: John Wiley, 1982), pp. 7–11; and George C. Beakley, *Careers in Engineering and Technology,* 3d ed. (New York: Macmillan, 1984), pp. 29–36.

23. James K. Feibleman, "Pure Science, Applied Science, and Technology: An Attempt at Definitions," *Technology and Culture* 2, no. 4 (Fall 1961), pp. 305–17.

24. David C. Gruender, "On Distinguishing Science and Technology," *Technology and Culture* 12, no. 3 (July 1971), p. 461.

25. There are, it may be noted, suggestive parallels between the distinctions drawn here and those to be found in the analytic philosophies of language, action, and mind. In the analysis of language, for instance, one finds discussion of a diversity of human linguistic acts: announcing, persuading, proposing, encouraging, promising, and so on. According to J. L. Austin, in *How to Do Things with Words* (New York: Oxford University Press, 1962), however, this diversity can be separated into perlocutionary and illocutionary linguistic acts, the former of which essentially involves the production of some effect (making?) and does not always require words (tools?), the latter of which does not necessarily produce any effect (using?) but does require a locutionary base (tools?). Furthermore, illocutionary acts (using?) can be a means to perlocutionary acts (making?), but not vice versa (at least not in the same way). Similar suggestive parallels can be drawn with analytic discussions of the relations between choosing (inventing?), deciding (designing?), and doing (making?).

26. Andrew Harrison, *Making and Thinking: A Study of Intelligent Activities* (Indianapolis, Ind.: Hackett, 1978), p. 1.

27. See Aristotle *Physics* 2.1. 193a12–17; *Politics* 7.17. 1337a2; and *Oeconomica* 1.1. 1343a26–b2.

28. Andrew G. Van Melsen, *Science and Technology* (Pittsburgh: Duquesne University Press, 1961), pp. 235–36.

29. Ibid., p. 236.

30. For a general analysis of skill see Patricia Benner, *From Novice to Expert: Excellence and Power in Clinical Nursing Practice* (New York: Addison-Wesley, 1984).

31. Claude Lévi-Strauss, *The Savage Mind* (Chicago: University of Chicago Press, 1966), p. 17.

32. Ibid., p. 19.

33. Bernard Rudofsky, *Architecture without Architects: A Short Introduction to Non-Pedigreed Architecture* (New York: Doubleday, 1964), opposite plate 4. (This is the exhibit catalog; pages are unnumbered.)

34. Cf. Donald A. Schon, *The Reflective Practitioner: How Professionals Think in Action* (New York: Basic Books, 1983), pp. 171–76.

35. "It is the function of the engineering and development department of a modern corporation to take equipment which it has been decided by management to manufacture, and to do the detailed study of the design and manufacturing process which is necessary if the device is to be produced cheaply and in volume, and it is to be free from minor defects under field operation." (Francis Russel Bichowsky, *Industrial Research* [1942; reprint, New York: Arno Press, 1972] p. 26.)

36. For a discussion of these and other aspects of engineering from the viewpoint of the engineer, see "Functions of Engineering," in Smith, *Engineering as a*

Career, 4th edition, pp. 118–66. Supplementary reference: A. W. Futrell, Jr., *Orientation to Engineering* (Columbus, Ohio: Charles E. Merrill, 1961), especially chapter 16, "Functions of Engineering."

37. Francis Bacon (1561–1626) is the first historical figure to argue explicitly and at length for the need to cultivate inventing. In so doing he distinguishes between those inventions which have been made on the basis of an appropriate understanding of nature and those which have been virtually independent of scientific knowledge—and, it could be added, method. The former are what today would be called science-based inventions; the latter, more traditional or evolutionary inventions. See, for example, Bacon's "Thoughts and Conclusions," trans. from the Latin in Benjamin Farrington, *The Philosophy of Francis Bacon* (Chicago: University of Chicago Press, 1966) pp. 90–91.

38. The older usage is apparent in Horace Kallen's article on "Innovation" in the *Encyclopedia of the Social Sciences;* the more specialized meaning takes over in Richard B. Nelson's much longer entry on the same topic in the *International Encyclopedia of the Social Sciences,* under the same term. For the most comprehensive study of this theme, see Melvin Kranzberg and Patrick Kelly, eds., *Technological Innovation: A Critical Review of Current Knowledge* (San Francisco: San Francisco Press, 1978).

39. See, for example, S. C. Gilfillan, *The Sociology of Invention* (1935; reprint, Cambridge: MIT Press, 1963). Two other primary sources for philosophical reflection on the nature and meaning of inventing are John Jewkes, David Sawers, and Richard Stillerman, *The Sources of Invention,* 2d ed. (New York: W. W. Norton, 1969); and H. Stafford Hatfield, *The Inventor and His World* (New York: Dutton, 1933). The most philosophical discussion is René Boirel, *Théorie générale de l'invention* (Paris: Presses Universitaires de France, 1961).

40. Alfred North Whitehead, *Science and the Modern World* (1925; reprint, New York: Free Press, 1967), p. 96.

41. See John B. Rae, "The Invention of Invention," in M. Kranzberg and C. Pursell, eds., *Technology in Western Civilization,* vol. 1 (New York: Oxford University Press, 1967), pp. 325–36; and Daniel J. Boorstin, "The Social Inventor: Inventing the Market," and "Communities of Inventors: Solutions in Search of Problems," in *The Americans: The Democratic Experience* (New York: Random House, 1973).

42. See especially the English translation, "Technology in Its Proper Sphere" (from *Philosophie der Technik,* 1927), in *Philosophy and Technology,* ed. C. Mitcham and R. Mackey (New York: Free Press, 1983), pp. 317–34.

43. Gilbert Simondon, *Du mode d'existence des objets techniques* (Paris: Aubier, 1958).

44. David Pye, *The Nature of Design* (New York: Reinhold, 1964), p. 19.

45. A selection of representative texts other than those already cited: Morris Asimov, *Introduction to Design* (Englewood Cliffs, N.J.: Prentice-Hall, 1962); J. R. Dixon, *Design Engineering: Inventiveness, Analysis, and Decision Making* (New York: McGraw-Hill, 1966); Myron Tribus, *Rational Descriptions, Decisions and Designs* (New York: Pergamon, 1969); Gerald Nadler, *The Planning and Design Approach* (New York: John Wiley, 1981); Vladimir Hubka, *Principles of Engineering Design* (London: Butterworth Scientific, 1982); G. Pahl and W. Beitz, *Engineering Design,* ed. Ken Wallace (London: Design Council; and Berlin: Springer-Verlag, 1984); and Michael J. French, *Conceptual Design for Engineers,* 2d ed. (London: Design Council; and Berlin: Springer-Verlag, 1985). The first chapter of Pahl and Beitz, translated from the German (of 1977), provides histor-

ical background on the devleopment of systematic design and debates about methods.

46. "Report on Engineering Design" (MIT Committee on Engineering Design), *Journal of Engineering Education* 51, no. 8 (April 1961), p. 647.

47. Woodson, *Introduction to Engineering Design*, p. 3. In the *McGraw-Hill Dictionary of Science and Technology* (1984) design is defined as "the act of conceiving and planning the structure and parameter values of a system, device, process, or work of art."

48. Pahl and Beitz, *Engineering Design*, p. 1.

49. See José Ortega y Gasset, "Thoughts on Technology," in *Philosophy and Technology*, ed. C. Mitcham and R. Mackey (New York: Free Press, 1983), pp. 295–97.

50. Joseph Edward Shigley, *Theory of Machines* (New York: McGraw-Hill, 1961), pp. v and vi.

51. E. F. O'Doherty, "Psychological Aspects of the Creative Act," in *Conference on Design Methods*, ed. J. Christopher Jones and D. G. Thornley (New York: Macmillan, 1963), pp. 197–203.

52. "Report on Engineering Design," *Journal of Engineering Education* 51, no. 8 (April 1961), pp. 647–48, n. 57.

53. For a related discussion of phantasmal thinking, see L. R. Rogers, "Sculptural Thinking—1" and "Sculptural Thinking—3" (part 2 is a commentary by Donal Brook), in *Aesthetics in the Modern World*, ed. Harold Osborne (New York: Weybright and Talley, 1968). For an empirical account of the high correlation between drawing and general engineering abilities, see Steve M. Slaby and Arthur L. Bigelow, "Engineering Graphics—A Predictor for Academic Performance in Engineering," *Journal of Engineering Education* 51, no. 7 (March 1961), pp. 581–87. At the conclusion of their statistical presentation, the authors suggest that "graphics is part of the thinking of an engineer" (p. 586), because "the ability to 'think' space is a necessary condition if we are to define engineering correctly" (p.587).

54. Edwin T. Layton, Jr., "Technology as Knowledge," *Technology and Culture* 15, no. 1 (January 1974), p. 36.

55. Ibid., pp. 37–38.

56. These factors are usually enumerated as four: materials, interrelation of parts, methods of construction, and effect of the whole upon those who will become involved with it. Only the first three, however, can actually be analyzed in the drawing itself; the fourth denotes a social context that is not amenable to quantification and is, in fact, a stumbling block and source of frustration to many engineers. For example, technically speaking, a long bridge can be constructed that is completely safe but that, if it is flat, because of the curvature of the earth will appear to an approaching motorist to be sagging; as a result the public will be afraid to use it. This compels the engineer to arch the bridge in a way that is not required by any of the first three factors. A second example: floors in concrete buildings must have almost twice as much concrete in them as they really need to support a designated load in order to keep them from vibrating in ways that pose no structural dangers but would make the occupants nervous.

57. For an elementary discussion of engineering modeling that unintentionally brings out its inherent character as miniature construction, see "Modeling," in *The Man-Made World: Engineering Concepts Curriculum Project* (New York: McGraw-Hill, 1971), pp. 139–78. "Models are used, not only to describe a set of ideas, but also to evaluate and to predict the behavior of systems before they are

built. This procedure can save enormous amounts of time and money. It can avoid
expensive failures and permit the best design to be found without the need for
construction of many versions of the real thing. Models evolve, and it is customary
to go through a process of making successive refinements to find a more suitable
model" (p. 177).

58. Jean-François Lyotard, *The Postmodern Condition: A Report on Knowledge*, trans. Geoff Bennington and Brian Massumi (Minneapolis: University of
Minnesota Press, 1984), p. 44.

59. Stanley R. Carpenter, "Alternative Technology and the Norm of Efficiency," in *Research in Philosophy and Technology*, 6:65–76; and Allen
Buchanan, *Ethics, Efficiency, and the Market* (Totowa, N.J.: Rowman & Al-
lanheld, 1985). See also *Encyclopedia of the Social Sciences*, s.v. "efficiency." See
especially p. 439: "[T]here is no such thing as efficiency in general or efficiency as
such—there are simply a multitude of particular kinds of efficiency. Actions and
procedures which are efficient when measured with one measuring stick may be
inefficient when measured with a different measuring stick."

60. Aristotle, *Physics* 2.3. 194b16–195b30.

61. It is interesting, however, that the original *Oxford English Dictionary* (1933)
did not include this meaning, which had to await the 1972 supplement for proper
recognition. On Smeaton's contribution to the development of this concept (if not
the actual term), see Arnold Pacey, *The Maze of Ingenuity: Ideas and Idealism in
the Development of Technology* (Cambridge: MIT Press, 1974), pp. 206ff.

62. Blaise Pascal, "De l'esprit géometrique et de l'art de persuader," where
there is a contrast with *l'esprit de finesse*. Cf. Plato's distinction between arts
based on numerical mensuration and those based on mensuration with regard to
the moderate, the fitting, the appropriate, the needful (*Statesman* 284e). This
reference is suggested by David Rapport Lachterman, *The Ethics of Geometry: A
Genealogy of Modernity* (New York: Routledge, 1989).

63. Plato, *Philebus* 55d–56c. For commentary, see Carl Mitcham, "Philosophy
and the History of Technology," in *The History and Philosophy of Technology*, ed.
George Bugliarello and Dean B. Doner (Urbana: University of Illinois Press,
1979), pp. 174–77.

64. The tendency to replace the word *efficiency* with *input-output analysis*
(although the concern is really with the ratio of output over input) is documented
by the two editions of the basic reference work in the social sciences. The
Encyclopedia of the Social Sciences has an article on the former, whereas in the
International Encyclopedia of the Social Sciences this is replaced by an entry on
the latter.

65. "Among the problems here are those concerning the lever. Indeed, it is
incredible that a larger weight can be moved by a weak power, even when more
weight is applied; for the same weight that a human cannot move without a lever,
one quickly moves by applying the weight of the lever." (Pseudo-Aristotle, "Mechanical Problems" 847b1–15.)

66. See, for example, Salomon de Caus, *Les raisons de forces movantes* (1615).

67. It is important to note that the efficiency of an economy is not the same
thing as economic efficiency. An economy can be described as efficient in production (when there is no way to increase outputs with available inputs), in exchange
(when it is not possible to alter the distribution of goods and services so as to leave
one or more persons better off without leaving anyone worse off), and in the
relation between production and exchange.

68. See, for example, William E. Akin, *Technocracy and the American Dream:*

The Technocracy Movement, 1900–1941 (Berkeley: University of California Press, 1977); and Howard P. Segal, "The Technological Utopians," in *Imagining Tomorrow: History, Technology, and the American Future,* ed. Joseph J. Corn (Cambridge: MIT Press, 1986), pp. 119–36.

69. On one aspect of this issue, see Alex C. Michalos, "Efficiency and Morality," *Journal of Value Theory* 6, no. 2 (Summer 1972), pp. 137–43. For an argument against the compatibility of input-output calculations and that religious morality which is constituted by the pursuit of spiritual perfection, see Dietrich von Hildebrand, "Efficiency and Holiness," in *The New Tower of Babel* (New York: J. P. Kennedy, 1953), pp. 205–43.

70. See Henryk Skolimowski, "The Structure of Thinking in Technology," *Technology and Culture* 7, no. 3 (Summer 1966): 371–83. Skolimowski was a student of the Polish praxiologist, Tadeusz Kotarbinski; see Kotarbinski's *Praxiology: An Introduction to the Sciences of Efficient Action* (Oxford: Pergamon Press, 1965).

71. Herbert A. Simon, *The Sciences of the Artificial,* 2d ed. (Cambridge: MIT Press, 1981), p. 135.

72. See Jacques Ellul, *The Technological Society,* trans. John Wilkinson (New York: Knopf, 1964), pp. xxv, 24, and 79.

73. For a good update on the spectrum of research regarding such input-output forms of rationality, see Jon Elster, ed., *Rational Choice* (New York: New York University Press, 1986).

74. Simon, *Sciences of the Artificial,* p. 130.

75. Ibid., p. 156.

76. Schon, *The Reflective Practitioner,* p. 175.

77. Billy Vaughn Koen, *Definition of the Engineering Method* (Washington, D.C.: American Society for Engineering Education, 1985).

78. I. C. Jarvie ("The Social Character of Technological Problems: Comments on Skolimowski's Paper," *Technology and Culture* 7, no. 3 [Summer 1966], pp. 384–90) objects that what the engineer strives for is really determined by the social definition of the problem. For instance, in warfare, there are times when civil engineers are called upon to design a bridge for speed of construction rather than durability. But Skolimowski's point is that within such historically and socially set parameters as materials, cost, and time a civil engineer qua civil engineer will always strive for as much durability as is feasible. In fact, Jarvie's own example tells against him, because it is the military rather than a civil engineer who would be called upon to design a pontoon bridge for maximum military efficiency (that is resistance to damage by firepower and mobility).

79. Although beyond the consideration of the present remarks, this representation process may be related to the creation of writing and what Bruno Latour calls "inscription devices." See Bruno Latour, "Visualization and Cognition: Thinking with Eyes and Hands," *Knowledge and Society: Studies in the Sociology of Culture Past and Present,* vol. 6 (Greenwich, Conn.: JAI Press, 1986), pp. 1–40. See also Ivan Illich and Barry Sanders, *ABC: The Alphabetization of the Popular Mind* (San Francisco: North Point Press, 1988).

80. Some approaches to philosophical discussion of design methodology beyond those already referenced are J. Christopher Jones and D. G. Thornley, eds., *Conference on Design Methods* (New York: Macmillan, 1963); Gerald Nadler, "An Investigation of Design Methodology," *Management Science* 13, no. 10 (June 1967), B-642–55; Bohdan Walentynowicz, "On Methodology of Engineering Design," in *Proceedings of the 14th International Congress of Philosophy,* vol. 2

(Vienna: Herder, 1968), pp. 586–90; Valdimir Hubka, ed., *Review of Design Methodology,* Proceedings of the International Conference on Engineering Design in Rome 1981 (Zurich: Heurista, 1981); and W. E. Eder, ed., *Proceedings of the 1987 International Conference on Engineering Design* (New York: American Society of Mechanical Engineers, 1987). Friedrich Rapp's *Contributions to a Philosophy of Technology* (Dordrecht: Reidel, 1974) includes two articles on design methodology by M. Asimov and R. J. McCrory.

81. Art and architecture books on the subject of design invariably concentrate on issues of form. A book on roof design, for example, provides an inventory of various ways to build roofs—not ways as actions, but ways as forms, patterns, shapes; one on lighting design contains pictures and drawings of various alternative formal solutions to lighting design problems. A work whose subtitle aptly illustrates this approach is Kurt Hoffmann, Helga Friese, and Walter Meyer-Bohe, *Designing Architectural Facades: An Ideas File for Architects* (New York: Whitney Library of Design, 1975). For some comprehensive architectural discussions of design that approach the philosophical, see Paul J. Grillo, *What Is Design?* (Chicago: P. Theobald, 1960); Christopher Alexander, *Notes on the Synthesis of Form* (Cambridge: Harvard University Press, 1964); David Pye, *The Nature of Design* (New York: Reinhold, 1964); Bryan Lawson, *How Designers Think: The Design Process Demystified* (London: Architectural Press, 1980); and Peter G. Rowe, *Design Thinking* (Cambridge: MIT Press, 1987).

82. Industrial design, especially the Bauhaus school of industrial design, is an attempt to bridge this gap between art and engineering, and either to include aesthetic formal properties in the design process or to find aesthetic value in purely functional designs. In fact, however, the attempt has led to the triumph of engineering efficiency as influenced by economic pressures. For a brief overview, see *New Encyclopaedia Britannica,* s.v. "industrial design."

83. Edward S. Casey's *Imagining: A Phenomenological Study* (Bloomington: Indiana University Press, 1976) provides some foundations for a more extended comparison of engineering and artistic imagination.

84. One of the more witty formulations of this thesis is Kranzberg's law that "technology is neither good nor bad, nor is it neutral." See Melvin Kranzberg, "Technology and History: Kranzberg's Laws," *Technology and Culture* 27, no. 3 (July 1986), pp. 544–60.

85. Emmanuel Mesthene, "How Technology Will Shape the Future," in *Philosophy and Technology,* ed. Mitcham and Mackey, pp. 116–17.

86. Steven Goldman, "The *Techne* of Philosophy and the Philosophy of Technology," in *Research in Philosophy and Technology* 7 : 124.

87. Ibid., p. 123.

88. Example: In frustration at the difficulties of communicating with some anti–nuclear power humanities students, an engineering professor was overheard to remark that "I have more in common with engineers in Russia than I do with these students."

89. An approach that is sometimes said to begin with Thomas Kuhn's *The Structure of Scientific Revolutions* (Chicago: University of Chicago Press, 1962) but has been given renewed vitality by, for example, Bruno Latour's *Science in Action: How to Follow Scientists and Engineers through Society* (Cambridge: Harvard University Press, 1987) and Joseph Rouse's *Knowledge and Power: Toward a Political Philosophy of Science* (Ithaca: Cornell University Press, 1987).

90. For a deft account of this movement, see John M. Staudenmaier, *Technology's Storytellers: Reweaving the Human Fabric* (Cambridge: MIT Press,

1985), especially chap. 3, "Science, Technology, and the Characteristics of Technological Knowledge," pp. 103–120.

91. See Carl Mitcham, "Responsibility and Technology: The Expanding Relationship," in *Technology and Responsibility*, ed. Paul T. Durbin (Dordrecht: Reidel, 1987), especially "Engineers, Professional Responsibility, and Ethics," pp. 13–18.

92. Lambert J. Van Poolen, "Technological Design: A Philosophical Perspective," in *Proceedings of the ASEE Annual Meeting* (American Society for Engineering Education, 1985), 5:13. See also Stephen V. Monsma, ed., *Responsible Technology: A Christian Perspective* (Grand Rapids, Mich.: Eerdmans, 1986). For a statement of the same position within a secular framework, see Victor Papanek, *Design for Human Scale* (New York: Van Nostrand Reinhold, 1983).

Part 2
Epistemological Issues

The Social Captivity of Engineering

STEVEN L. GOLDMAN

Two characterizations of modern engineering—as applied science and as the primary agent of technological change—are nearly ubiquitous, yet they fundamentally misrepresent its theory and its practice. Concurrently, they obscure both the characteristic intellectual problems that engineering poses and the actual relations of engineering to American society. My objective in this essay is to concentrate on the social relations of the practice of engineering, deferring to a second essay[1] discussion of the intellectual problems posed by its theory.

In the end, I wish to argue that engineering is today captive to society, and in two ways. Intellectually, engineering is captive to a cultural prejudice that denies the very existence of a theory of engineering—that is, of a distinctive conceptual framework, a *theoria,* or perspective on the world, of engineering's own—by reducing engineering to devising applications of the products of scientific theorizing. The consequent theoretical transparency of engineering provides a rationale for concluding, quite incorrectly, that all of the "serious" intellectual problems—epistemological, metaphysical, moral, sociological—attach to science, the principles of whose practice are supposed to comprehend the practice of engineering as well. At the same time, engineering practice itself is captive. It is captive to social determinants of technological action that selectively exploit engineering expertise, define the problems engineers are to address as well as the terms of acceptable solutions, and shape the technical knowledge available at any time by influencing educational curricula and the direction of research and development initiatives.

These are not two separate captivities. They are complementary expressions of a fundamental usurpation of the intellectual and social dimensions, respectively, of engineering as an autonomous discipline. The intellectual subordination of engineering to science anchors the subordination of engineering practice to social determinants of technological action, while the fact of the latter subordination reinforces the legitimacy of the former.

The belief that engineering drives technology disguises the actual subordination of engineering to the institutional dynamics of technological action. Technological action is a social process in which engineers participate rather than something that engineers do. The forces driving that process are initiated, and constantly modulated, by context-specific, hierarchically structured "managerial" decisions that selectively apply technical knowledge to the accomplishment of objectives characteristic of the institutions those decision makers serve. The term *managerial* is used to emphasize the extent to which technological action is determined by parochial considerations of a highly arbitrary character. These considerations are interpretations by people with decision-making authority—in a company, corporation, venture capital firm, or government agency—of what the exploitation of technical knowledge in one way rather than another can achieve for them, given the institutional, personal, and societal influences to which they consciously and unconsciously respond.[2]

Technology is thus a decision-dominated, rather than a (technical) knowledge-dominated process. Furthermore, even when the decision makers are engineers by training, the decisions they make reflect their place in the managerial hierarchy. Neither objective natural limitations nor what engineers and scientists know determine the kinds of machines that are built and the physical characteristics of those machines. That is determined by the managerial agendas of the institutions to which engineers make their knowledge available, together with the knowledge those engineers possess, knowledge that also reflects the influence of those agendas.

Illustrations of this abound. That, with one exception, all of the nuclear power plants in the United States are light water reactors and of a de facto standard size (since the late 1960s) of 1000–1200 megawatts is not at all a reflection of the only reactors that American nuclear engineers could have built in the 1960s, or would have built had the choice been theirs. The dominance of light water reactors is overwhelmingly a consequence of the involvement of General Electric and Westinghouse in providing reactors for nuclear submarines, as a result of which they accumulated considerable experience with light water reactors and acquired facilities that could readily be turned to producing commercial power reactors of the same sort.[3] Similarly, the size of reactors built after the mid-1960s is a consequence of GE and Westinghouse having offered 500–600 megawatt reactors to utilities in 1963–65 at low prices, on a fixed cost, turnkey basis in the hope of inducing enough utilities to "go nuclear" to create a commercial nuclear power industry.[4]

The very existence of a commercial nuclear power industry is a consequence not of the nuclear science and engineering knowledge bases available in the 1950s, and not of a perceived need on the part of utilities for alternatives to coal and oil, but of deliberate federal policy decisions. For a variety of national security, political, and economic reasons, the Truman and Eisenhower administrations, the AEC, and the Joint Committee on Atomic Energy were committed to bringing civilian nuclear technologies into the marketplace.[5] The AEC, in addition, shaped the available nuclear knowledge base by the types of reactor and nuclear weapons research and development projects they funded, the production and testing processes they supported, the safety and reliability studies they required, and the standards they set. Nuclear science and engineering, therefore, did not constitute the basis for the introduction of commercial nuclear power plants, for example, by offering a technology that was a demonstrably cost-effective alternative to existing electric power technologies; nor did they determine either the type or the size of reactors subsequently constructed.[6] All of these followed from managerial agendas. Furthermore, the content of nuclear science and engineering was significantly affected by the same decisions.

The fusion power program in the United States has followed a similar course, as Joan Bromberg has documented.[7] The commitment to federally funded fusion power development and the distribution of support among magnetic confinement fusion reactors and inertial confinement reactors; among research, design studies, and construction of experimental and prototype machines of different types; among basic reactor science (mostly physics) and reactor engineering has been influenced significantly by political and military considerations.

The computer industry offers a number of homelier illustrations that are nevertheless of greater social significance (so far) than nuclear power or fusion. The pervasiveness of the computer in American society today is substantially the result of the invention of personal computers in the late 1970s, and of the introduction of the IBM PC in the fall of 1981. To an astonishing degree, this one machine transformed the market for personal computers, enormously expanding it and accelerating the spread of computer technologies and computerization throughout the workplace, in schools and homes, and in the consciousness of society. In the process, the

IBM PC earned billions of dollars for IBM and billions more for the manufacturers of PC "clones," yet the IBM PC was not a technically advanced computer. The corporate market especially had remained aloof from personal computers before 1982, at least in part because the manufacturers were new, small, and unstable as a result of the intense economic and technical competition that made machines obsolete that had been state of the art only months earlier. No long-term standard existed on which corporate planning could rely.

IBM's PC changed all that. It represented a radical policy departure for IBM management. Many of the components were manufactured by other companies and only assembled, along with IBM-manufactured components, into IBM-labeled machines. The system software was not written by IBM but was commissioned from a then-small company that as a result of that one contract has become an industry giant, Microsoft. The computer's "architecture," or fundamental design, was open, encouraging the creation of a third-party industry providing optional non-IBM hardware to plug into the PC and non-IBM software to run on it. Seven years later, the IBM PC–type of computer is still the dominant personal computer, and the many millions of them in daily use, supported by literally thousands of companies providing PC-based products, suggest that it will survive a lot longer, in spite of the constant, rapid advance of computer technology. (As a matter of fact, the Apple II, introduced in 1977 and updated but not fundamentally changed, for the very same reasons continues to sell very well and to generate perhaps as much as half of Apple revenues even though its successor, the Macintosh, is an incomparably superior machine by certain technical performance criteria.)

The success of the Apple computer corporation, as distinct from the Apple computer itself, was made possible by the management of Hewlett-Packard having turned down the offer of the computer's designer, Steve Wozniak, at the time a Hewlett-Packard employee, for Hewlett-Packard to develop the machine for production.[8] That the IBM PC precipitated the wholesale adoption of personal computers in business was made possible in part by the decision of Digital Equipment Corporation's president Ken Olson not to support production of a personal computer based on the very successful DEC PDP-8, the basic model of which the PDP-8/F sold for $6,000 in 1977. Olson's response to a proposal for scaling down the PDP-11 was that he could think of no reason why people would want to own their own computer![9] Tracy Kidder's *The Soul of a New Machine* documents the creation of a new model of minicomputer at Data General Corporation under very specific managerially imposed design constraints.[10]

The commitment to the Apollo Program was a totally political commitment, a fact that has had enormous consequences for the course of American space policy, space science, and aerospace engineering from the 1960s to the present.[11] The technical design of the space shuttle was dramatically affected by the political compromises with Congress that NASA had to make in order to win funding for it. The design of the space station is today undergoing similar modification, with consequences we cannot now foresee, as NASA and the outgoing Reagan administration wrestled with Congress over the level of funding to be assigned to the project, which has seemingly no other rationale for being built than to win national prestige.[12]

The defense industry provides too many illustrations of the dominance of managerial decision making over technical knowledge to even list. The most recent of them include the characteristics of the B-1 bomber, which reflect the political maneuvering, especially by the air force and Rockwell Corporation, necessary to build the plane in the face of persistent opposition for twenty years by presidents of both parties and many members of Congress. In the process, changes to the design specifications made in response to budgetary and political compromises, as well as advances in a host of relevant technologies, have resulted in a 100-plane fleet currently rated as operational but unable to fulfill its mission.[13] Similarly, the technical characteristics of the Los Angeles class nuclear submarines reflect decisions from above—managerial decisions as I use the term here—that constrained technical development and are reflected in the performance of these submarines.[14] Their weight especially, which limits speed and affects their mission, is the consequence of a decision not to use titanium hulls, as the Russians do, for the submarines that the Los Angeles class is supposed to be able to track and hunt down. As a result of their speed deficit, and perhaps an operational depth deficit as well, research and development of collateral technologies for finding and destroying Russian submarines has had to be pursued, which, in the absence of these deficits, would have been unnecessary.[15] Finally, the Strategic Defense Initiative, in which former President Reagan took an active role from the beginning, has seen the course of its technical development, the nature of its mission, and the technical specifications in its contracted R&D defined and redefined on the basis of political and managerial objectives.[16]

The implications of the characterization of technological innovation as a managerial decision–dominated, rather than a technical knowledge–dominated process are obvious. The dynamics of technological action are, in the first instance, to be located in relevant institutional contexts, not in engineering or scientific expertise. Overwhelmingly, today, these contexts are of two types: large corporations and federal governmental agencies. One-sixth of all U.S. scientists and engineers employed as such work in the more than 600 federal laboratories. One-third of all the scientists and engineers in the United States are concentrated in the defense industry. Approximately 80 percent of U.S. engineers are employed in private industry, and in 1969, 70 percent of them were employed by just 1 percent of the firms that employed engineers. While the individual companies making up that 1 percent have changed, this concentration very likely continues today.[17] Most engineers in industry are employed in a small number of areas, each of which tends to be dominated by a small number of very large corporations: computers and electronics (including communications), aerospace, chemicals, engineering services, energy, transportation. The practice of engineering is therefore typically embedded within the activities of large-scale enterprises characterized by hierarchical managerial structures and employing hundreds, often thousands, of engineers. The function of management in these enterprises, no matter how high the level of technology, is not to serve the engineering branch, but the reverse: to bring the expertise of the engineering branch to bear on the execution of objectives set by management.

The managerial structure of an organization constitutes the armature around which are wound the means available to its executive officers for providing them with information, for realizing their objectives, and for monitoring the organization's operations. The means—manufacturing, R&D, finance, sales, marketing—seem to be the active elements in the organization, while management reacts to their activity and coordinates it. In fact, however, management is the actor in all these operations, the prime mover instigating all these operations as reactions to its policy agenda. Some of these operations are directly commanded, but the majority by far are only signified by upper-level policies with broad specifications. It becomes the charge of lower levels of management actively to assimilate these objectives and to effect their realization by giving them progressively more detailed levels of specification, drawing upon the particular production skills and engineering expertise available.

A number of observations need to be made about these remarks. First, they apply to government agency operations—that is, to the subordination of engineers to their managers at the Environmental

Protection Agency, the National Bureau of Standards, and the Nuclear Regulatory Commission—as well as to private industry. Moreover, within private industry, they apply to start-up companies as well as to the largest corporations. That is why the term *manager,* though used somewhat idiosyncratically here, is an accurate descriptor. Second, this account of the managerial function applies as much to engineers in mainstem managerial positions, and even to engineers managing the engineering function within a company, as to all other managers. For engineers must expand, some would say override, their professional sensibilities as engineers in order to assimilate the managerial agenda.

Managers do not simply do what they are told by their superiors. The management of a large company would be almost impossibly unwieldy if that were the case. Managers need to adopt as their own the objectives set by their superiors and to wield the resources given to them on behalf of accomplishing those objectives. Sometimes this requires setting aside what they might have thought of as prudent engineering had they still been engineers, instead giving priority to cost considerations, scheduling conflicts, performance commitments, or stylistic requirements. On the night before the doomed space shuttle *Challenger* launch, during the teleconference about launch approval between Morton Thiokol engineers and managers and mid-level NASA officials with launch authority, the Thiokol engineers (with the NASA people off-line temporarily) were asked by their superior to take off their engineering "hats" and to put on their management "hats." The former were causing them to oppose the launch because of concerns about the projected launch-time temperature and solid-fuel rocket-booster seal performance. The latter prompted them to take into consideration their client's (NASA's) wishes and the possible impact of disappointing the client on future Thiokol-client relations, including pending booster-rocket contract renewals and new contract proposals.[18]

Tracy Kidder's portrait of Tom West in *Soul of a New Machine* shows an engineer who clearly aimed to design a computer far superior to the highly successful "Eagle"/Eclipse MV-6000 his team ultimately produced. But as a mid-level engineering manager with aspirations to upper-level management, he accepted Data General President de Castro's design constraints and imposed them on the engineers working for him. Some of the senior engineers under West also preferred a superior design, and it was West's job to cajole them into accepting de Castro's constraints and to develop a machine that they considered to barely tap the kind of engineering creativity they were capable of.[19]

Kidder's account of West's performance as an engineer-manager with higher managerial ambitions matches perfectly the description given by Walter Vincenti of the development of engineering specifications within a company. While the design process is different in different branches of engineering, innovation in any field of technology "depends on the relatively unstructured conceptual activity" of high-level managers translating "often ill-defined commercial or military needs into a concrete technical problem for the level below." At the lower levels, engineering problems "are normally well defined and activity tends to be highly structured."[20] The specification process thus parallels the hierarchical structure of management.

The logic of any given technological action decision, then, is specific to its institutional context. That is, each decision, from the grandest commitment to build a reusable space vehicle to the selection of one particular coating material to protect from corrosion the stainless steel of the solid-fuel booster-rocket shells makes sense only in relation to its institutional context at the time that it was made.[21] The desire to attribute to technical expertise primary responsibility for the products associated with technological action—that is, for artifacts, processes, and techniques—is understandable but misleading. It has all the appeal of substituting simple appearances for a complex reality. Its popularity reinforces the power and the authority of the primary technological actors, namely, the managerial decision makers, by obscuring the value-laden nature of their role. It is the selectivity of the application of technical knowledge that determines the properties of technological action, and this selectivity is a managerial prerogative. Appreciating that the locus of selectivity is contingent managerial decision making rather than objective limits of technical knowledge—or even objective performance standards or market forces, as David Noble has recently argued[22]—decisively undermines popular notions of technology and technology management as propelled by objective forces.

In fact, however, there is no general recognition that the dynamics of technological action lie in the arbitrary policy agendas of institutional contexts rather than in technical knowledge bases. As a result, neither is there a general recognition that the ways in which engineers are utilized within the institutions that employ them can be at least as significant for what engineers *do,* as what they *know.* Personnel utilization policies are, of course, a managerial agenda item. So is, to a certain degree, what it is that a company's engineers know. Both are consequences of whom managers decide to hire, which lines of internal R&D they fund, which external R&D or patents they buy or license.[23] Paradoxically, the engineering community supports the public's misperception of technology, thereby acquiescing in its own social captivity.

Why this should happen is not clear, but all organized attempts to escape from this captivity and assert the independence of engineering have failed. I refer to the efforts of Morris Cooke and others to wrest control of the engineering professional societies from industry domination in the period 1915–20, the overlapping and equally abortive technocracy movement supported by Frederick W. Taylor, Cooke, Henry Gantt, Thorstein Veblen, and Gifford Pinchot, among many others, and various attempts at organizing engineering unions through the mid-1950s.[24] As a matter of fact, essentially from the birth of modern engineering in the United States after the Civil War to the present, American engineers—and American engineering undergraduates—have identified closely with the goals and the policies of their corporate employers.

Edwin Layton and Samuel Florman, who otherwise have very different conceptions of engineering and the engineering-technology-society relationship, agree on this point. "Corporate culture does not consist of 'good' engineers seeking to protect the public from 'bad' managers. . . . Engineers who work for corporations by and large identify with the goals of these corporations and simply do not feel an urge to be freed to protect the public interest."[25] For Layton, conformity to the interests of their employers is built into engineering education,[26] and Bruce Sinclair has described in detail a play satirizing this conformity that was written and performed for the public by the Engineers Club of St. Louis in 1930 on the fiftieth anniversary of the American Society of Mechanical Engineers.[27]

Every Engineer: An Immorality Play traces the career of a young engineer from his idealistic beginnings to his bitter retirement. The new graduate feels instantly loyal to his first employer and remains so for many years, even though he comes to realize that his employer is motivated primarily by profit, is manipulative and exploitative of his employees, is addicted to financial gimmickry, and is disinterested in any but the company's best interests. Finally the no-longer-young man quits and sets up as a consultant, only to discover that technical ingenuity must be compromised if he is to get contracts from the business and political leaders who control them.

This submissiveness of engineers to the managerial hierarchy is puzzling in light of the recognition early in the twentieth century that the then-new corporate managerial structure was itself a new technology and that it was the central technology of the industrial age. Within this structure, as the British economist Alfred Marshall wrote, the technical experts in engineering and finance certainly hold sway, but they have not translated this dependency on their expertise into corporate power.[28] In the United States, fewer than a third of industrial chief executive officers, and only a quarter of corporate board members, have technical backgrounds in science or engineering. (The percentages are just the reverse in Japan, and in Western Europe they are nearly so.) J. K. Galbraith, in *The New Industrial State,* expanded upon Marshall's thesis, insisting on the power of the technocracy because of the growing dependency of corporate well-being on planning, and the dependency of planning on the expert knowledge possessed by technocrats.[29] But however inevitable its existence may seem to be in theory, no such power is evident in practice.

Seventy years after Marshall wrote, and twenty years after *The New Industrial State,* engineers are more than ever entrenched as executors of the policies, and therefore of the values, of their employers. Even the flurry of interest in engineering ethics that developed in the wake of the 1960s attacks on technology, and on engineering as responsible for technology, has largely restricted itself to issues of detail rather than policy. As Langdon Winner once put it, only a little facetiously: most engineering ethics scenarios take the form of wrestling with the question of whether an engineer has an obligation to warn workers painting the nose cones of nuclear-armed missiles if he or she learns that the paint's fumes are toxic![30] With few exceptions, the whistle-blowing episodes that have attracted the greatest media coverage are analogous. That is, they are instances in which engineers have objected to managerial decisions regarding particular engineering practices—the case of the BART engineers, for example—while leaving fundamental policies uncriticized. A rare example of the latter is the effort to rally the nation's scientists and engineers, particularly systems software engineers, to refuse on professional grounds to participate in SDI projects.

The upshot of all this is that engineers in practice assimilate the values embedded in the policies and decisions of their employers relating to how engineering is to be used to achieve managerial objectives. Public interest is a vacuous criterion for engineering practice, not because it does not exist or ought not to be weighed, but because no means exist for bringing it to bear on the decision-making process that controls the expression of engineering knowledge. (If such means did exist, the problem of determining what the public interest is would suddenly become pressing!) But the managers themselves serve the agendas their institutions adopt, and so the setting of managerial agendas for technological action becomes a central factor bearing on the conduct of engineering.

It would be impossible to list all the reasons why managers make the decisions they do. Speaking generally, however, the prevailing relationships between society and the institutions involved in technological action are projected onto engineering and define the character of engineering's subordination to management. These relationships include the bases of the value judgments that managers make in setting policies and the means that will be acceptable for executing them: acceptable in that social setting at that time. Today, for example, businesses have to take into account the pollution of natural resources that their operations cause, the long-term safety of working conditions, and the impact of their facilities on the surrounding social settings, to a far greater degree than ever before.

The form of the technology-society relationship is thus of considerable importance for understanding why technologies develop in the ways that they do, at the times that they do, and have the impacts that they do. It is important as well for understanding how to intervene effectively in the innovation process, how best to regulate technological activity, and how to modify its social and physical impact. It is also central to understanding engineering, because the "boundary conditions" of engineering problem solving derive from managerial interpretations of what is worth doing at that time, for that institution, and how it ought to be done.

The first generation of modern historians of technology included a number of thinkers—Abbott Payson Ussher, Sigfried Giedion, Lewis Mumford—who were deeply committed to the notion that technologies developed not out of an objective technological imperative, but because of forces deriving from their social relations.[31] Of these, Mumford, whose interpretations of history echo themes in G. B. Vico's *New Science,* has developed the most detailed philosophy of technology. In *Technics and Civilization* and in the two parts of *The Myth of the Machine,* Mumford locates what he perceives as the pathology of technological innovation in the West since the Renaissance, in a divergence between the symbol-generating activity of human consciousness and technological activity.[32] The latter is firmly under the control of institutions and individuals who use technology to generate and maintain power, wealth, and privilege, sharing these to a calculated extent with various strata of society. The former needs to find ways of life that answer to its symbolic constructions, which are creative projects based on human interaction with the world (as in Vico). But that world is a product of the technologies we emplace in it, so that our creative symbolic activity is impoverished by the implementation of "inauthentic" technologies. The divergence between symbolic and technological activities has two consequences. Technological activity is increasingly destructive because, divorced from human symbolic activity, it feeds primarily on greed and power. Meanwhile, people feel increasingly unsatisfied with lives that are supposedly improving, because the ways people live reflect the technologies in place rather than aesthetic projections of human life possibilities.

Jacques Ellul, also revealing the influence of Vico, as well as of G. W. Hegel and Ernst Cassirer, has formulated a philosophy of technology as global as Mumford's.[33] Both men are committed to the view that technology today is destructive of human potential and an obstacle to human fulfillment. Both also attribute the failings of technology not to particular machines that should have been made differently, but to the form of technology's socialization. Only the system as a whole can usefully be blamed, which is the point of Ellul's frustratingly imprecise notion of *la technique*. For Ellul and Mumford, a reform of engineering is barely relevant to a progressive reform of technology. Winner's anecdote about engineering ethics is apposite here.

Only if engineers withdrew from current institutionalizations of technological action in favor of "better" ones could their behavior significantly affect the course and consequences of technological action. But, of course, such a revolutionary transformation could also be accomplished through political action, independent of activism on the part of engineers. In that case, the engineers would simply be given a new professional mandate, which, say, for Winner, would include taking account of the public interest as appropriately explicated. This would not be enough for Mumford. The social process of technology would have to be restructured in a way such that only technological activity in harmony with fertile symbolic life structures would be permissible.

Most recent students of the social relations of technology have focused much more narrowly than Mumford, Ellul, and Winner on the social determinants of innovation. Of these, Hugh Aitken, Edward Constant, Thomas Hughes, David Noble, and Bruno Latour seem to offer especially noteworthy models of this process. Constant locates the dynamic element of the innovation process in the sphere of the technical.[34] Someone, generally but not necessarily an engineer or scientist, projects performance limitations of a current technology and invents or initiates the pursuit of a new technology to circumvent those limitations. Naturally, the social environment strongly affects the rate at which this new technology is introduced, even determining whether it *is* introduced. Other determinants include the inventor-prophet's institutional connections, talent, personal magnetism, and commitment to realizing his or her vision. Constant chose the development of turbojet aircraft engines as an illustration of his thesis, but many others come to mind, from rocket-propelled vehicles in the generation of Robert Goddard to the introduction of 16-bit personal computers long before nontechnical users had begun to experience performance limitations in the existing 8-bit machines.

But while "presumptive performance anomalies" constitute one route to innovations, there are other routes as well. Not infrequently, an invention precedes a clear vision of what can be done with it, or an invention intended for one line of technological development makes a more powerful social impact in another line.[35] Sometimes, technological innovations are introduced primarily to aggrandize or protect certain companies and government agencies: the air force persistence in the matter of the B-1 bomber is an apposite illustration of the latter. And one can hardly define the personal computer as the response to someone's vision of improving existing information processing tasks. (Indeed, this may not be true even of the mainframe computers of the early 1950s.) It seems pretty clear that when personal computers were invented they had no clear purpose. To the extent that they were invented for a reason, it was to satisfy what amounts to a sophisticated tinkering "itch" among technical enthusiasts involved with the long-established hobby, and business, of personal electronic projects and kit-building.

There was thus an autonomous technical side to the introduction of personal computers. This incorporated the focused enthusiasm of individuals specifically interested in experimenting with computing and/or with the newly introduced integrated circuit and microchip technologies, together with the unfocused, generic enthusiasm for electronics of a vast hobbyist community, one segment alone of which—amateur radio operators—numbered several hundred thousand active members in the late 1970s. And there was a commercial side, representing the interests of companies offering electronic kits and supplies as well as companies servicing the hobby electronics and amateur radio communities, especially publishing companies, initially. What is not clear is the basis for the extraordinary public enthusiasm for computers, initially centered on games, then, briefly, on learning to program, and finally stabilizing around word processing, spreadsheet, and data base programs. The detailed course of development of hardware and software, from computer chip architectures to operating systems, programming languages and user interfaces, reflects a progressive shift of influence from technical enthusiasm to characteristically managerial concerns.[36]

In *Networks of Power*, Thomas Hughes emphasizes the significance of the resolution of "reverse salients," that is, of specific technical problems that appear during implementation of a grand innovation.[37] In describing the triumph of alternating over direct current electrical systems at the turn of the century, he points to the development of polyphase alternating current motors as an example.[38] Nevertheless, what Hughes's study above all communicates to the reader is the significance of the social context for the course of the technical and commercial development of a new technology, for the form of its implementation, and for the character of its social impact. "Technical problems," Hughes writes, "are sometimes in essence institutional and value conflicts," such that technological change cannot be explained by referring only to the technical. The "style of the various [electric power] systems," that is, the blend of technical knowledge and social influence peculiar to a particular time and place is essential to such explanation.[39]

In Chicago, electric power technology took precedence over politics, largely because Samuel Insull was successful in his political campaign to have the regional electric utility regulated by state rather than by local governments, and also because Insull made a firm commitment to the largest scale of generating equipment then available, even though the first several of the giant steam turbines he ordered did not perform as projected. In London, political and financial maneuvering dominated the technology so that the electrification of London, and of England, lagged behind Chicago and Berlin until well after World War I. In Berlin, politics and technical development were coordinated. The establishment of a Berlin city franchise that adopted the universal polyphase alternating current system as a standard sped up maturation of alternating current electrical technology. So did a national commitment to engineering education as a means of rapidly acquiring, commercially developing, and profiting from innovations, especially those invented elsewhere.

Because of the significance of the form of the technology-society relationship for the course of technological development, the "evolving power systems were not . . . driverless vehicles carrying society to destinations unknown and perhaps undesired."[40] Hughes's study thus reinforces the notion that factors explicitly external to the technical give control and direction to the evolution of the technical and determine the form that technical applications take. I have been arguing that, implicitly, the sphere of the technical already contains the mark of these "external factors," so that they are not truly external to it. In engineering, the technical is only superficially independent of the social. Society drives technological change, generically, through the struggles of new and existing vested interests within the framework of the prevailing political, economic, legal, and cultural institutions and values; specifically, such change comes about through what I have depicted as the mechanism of managerial activism in which these struggles find concrete expression.

David Noble's *Forces of Production* further reinforces this account of the interpenetration of the technical and the social.[41] Noble attempts to identify the dynamics at work behind the introduction of, first, numerical control and then computer numerical control (CNC) machine tools in the United States. Noble is particularly interested in why one particular form of numerical-control machine-tool technology, rather than another available and apparently equally viable one (record-playback technology), today dominates the field. For Noble, the dynamics of innovation lie in the institutionalization of political and economic power, as a result of which control, especially control of the workplace and the work force, is a central theme shaping the forms in which technical knowledge is applied to production. This thesis, perhaps like Constant's, may not have been intended to encompass all innovation, but it does encompass the management of all innovation that becomes socially significant, leading to the crystallization of vested political and economic interests, as well as those innovations that de novo implicate existing patterns of control, as the operation of machine tools does.

Noble's interpretation of control in the context of machine-tool technologies centers on corporate management's assertion of control over workers as a point of principle—as a point of class principle—virtually regardless of economic cost or impact on production efficiency. The imposition of new machine-tool technologies is seen as part of a continuing effort by management to reduce the power of labor relative to management by reducing the skill levels necessary to perform any given task. In the case of CNC machine tools, the air force provided the occasion for developing the CNC technology (volume production of extremely complex machined surfaces) and the funding. Defense contractors and academic engineers and scientists, primarily at MIT, collaborated with the contracting agencies, the former with the ulterior motive of using this occasion to develop a new means for advancing managerial control, the latter as enthusiastic allies, at least in part because the technology preferred by management was more complex technically than record-play-back and would make the implementation of the new technology more dependent on their own expertise.

On the basis of his earlier book, *America by Design*,[42] one can conclude that Noble sees CNC technology, and the convergence of military, academic and corporate interests upon it, as paradigmatic of technological innovation in the United States, not merely as an interesting, perhaps anomalous, instance of innovation. Technology as a social process is firmly in the grasp of an established group of politically and economically powerful institutions, served by a well-defined class of individuals who shape the process to defend and extend their own interests. This is only one interpretation of control as a dynamical feature of technological action. It is an extension of Harry Braverman's thesis in *Labor and Monopoly Capitalism*, and has other defenders, most recently Harley Shaiken.[43]

Many other writers have recognized control as central to the development and introduction of technologies. Jim Beniger has written a history of Western technology from the perspective of control, *The Control Revolution*.[44] Taylorism was explicitly centered on control issues, though the engineer and not management was to be the ultimate locus of control for Frederick Taylor (after whom the movement is named). This very likely explains the rapid loss of enthusiasm for Taylor's reforms by management. As already mentioned, Alfred Marshall pioneered the recognition of the new corporate managerial structures as control technologies, and engineering control of innovation in the name of rationality, efficiency, and social progress for the masses was a theme common to Morris Cooke, Henry Gantt, Thorstein Veblen, and the technocracy movement. A measure of public control over technology was central to Gifford Pinchot's efforts on behalf of the Giant Power bill in the 1920s, as it was to the battle over the creation of the TVA in the 1930s.[45] Many of the critiques of technology in the 1960s and early 1970s called for control of technology by the people, for the people, if only because "the people" (and the environment) are necessarily implicated in all technological action.[46] Even the (corporate and governmental) technology establishment was affected by this in the 1960s, as it attempted to assimilate the new agenda item of bringing engineering resources to bear on social problems, in addition to the traditional problems of production and national security. Finally, radical sociologists of science and technology, adherents of the so-called Strong Program, interpret control from the perspective of the community of scientists and engineers whose norms and patterns of interaction constrain and mold the professional behavior and particularly the reasoning of individual practitioners.[47]

Trevor Pinch and Wiebe Bijker have outlined a program for transposing to the analysis of artifacts contemporary sociological analyses of science.[48] The result is a clarification of the essential way in which research, development, and production processes are social constructions whose concrete products reflect largely unarticulated social mechanisms of control. Bruno Latour explicates these in his recent book *Science in Action,* taking as an illustration the invention, development, and marketing of the diesel engine.[49] The transformation of Rudolf Diesel's idea for a new type of engine, an invention certified as such by the award of a patent, into a commercial product is revealed to be a complex social process. This process was dependent on the crystallization of alliances of interests among a range of parties whose mutually negotiated involvements in the transformation soon became quite independent of the inventor. These parties determined that there would be an engine, decided upon its physical characteristics, and even gave it its name, the *diesel* engine. Diesel himself had long since lost control of the process. Control of the research, development, and production activities is what defines the end-product, so to understand technology it is necessary to uncover these controlling influences. It can never be enough to know the laws of nature and the available base of technical knowledge.

The identification of control as an elementary feature of technology by so many groups, with such varied intellectual and political agendas, reinforces the judgment that technological action is manipulated by various expressions of managerial willfulness that are fundamental to technological action. The account of technology that seems to me to capture most fully the richness of the complex interactions that converge on the innovation process is that of Hugh Aitken, particularly as developed in his most recent book, *The Continuous Wave*.[50]

As with Noble's CNC machine-tool technology and Hughes's electric power networks, and as I have argued elsewhere,[51] continuous wave (CW) broadcasting technology has a symbolic dimension to it that mutely signifies the special interests behinds its creation. The task of the historian, or the sociologist, or the philosopher, of technology is to make technologies declare the contingencies underlying their existence. In the case of CW broadcasting, these included two decades of competition among three distinct types of broadcasting technologies: spark, vacuum tube, and high-frequency alternator transmitters to be used in conjunction with a range of types of receivers. Also included were the inventions, efforts, and promotions of independent inventors, company engineers, and entrepreneurs, among them Lee de Forest, Reginald Fessenden, Cyril Elwell and Leonard Fuller, and Ernst Alexanderson. This competition was inseparably linked to the active intervention of rival policy commitments and interests of AT&T, GE, Marconi International (especially the American and British branches), the U.S. Navy, and the Royal Navy, along with the American, British, French, and German governments and the American and British financial communities. The public took a role as "consumer" of national security, as a source of funding for new facilities and for profit through financial manipulations constrained only by the prevailing legal climate regulating securities and banking activity, and as the target of commercial broadcasting schemes following the start-up of station KDKA in Pittsburgh in 1920.

In *The Continuous Wave*, Aitken undertakes to do justice to one episode of technological innovation by encompassing in his account of it all of the primary, mutually interrelated factors impinging upon it. The result is a convincing depiction of technology as a social process dependent upon, but certainly not determined by technical knowledge. Aitken's study shows how technical knowledge itself develops in response to funding and marketing opportunities, actual and anticipated, which can in turn implicate governmental and private business policies—there to be exploited or to be made the objects of lobbying efforts to change, replace, or put in place—as well as in response to personal enthusiasms of inventors, engineers, entrepreneurs, executives, politicians, and government officials.

To return to my main point, it is simply impossible, even in principle, to isolate the technical dimension of engineering from the concrete social circumstances of its expression in the form of actual, projected, and serendipitous applications. Not only is engineering an intentional activity whose object is socially defined in essentially arbitrary and willful ways, but *engineering is itself the object of an intentional activity, namely, managerial activity.* Appreciation of this concept must alter our intellectual apprehension of engineering. Although it is in thrall to the institutions that control its practice and is treated with condescension by the intellectual community, engineering is an extremely complex phenomenon that deserves to be studied in its own right.

It is much more complex than science, whose propositional content is intimidating (in large measure because of the symbolic language in which it is expressed) but whose logical structure and social relations are relatively simple. Engineering cannot be separated from the willfulness that pervades the determination of its object, whereas the Western intellectual tradition has embraced a particular notion of rationality that specifically excludes volition. Engineering cannot be separated from the actuality of its context, which is again problematic for Western philosophy, which since its Greek origins has embraced universality at the expense of contextual actuality. Engineering cannot be separated from a form of rationality that incorporates non-uniqueness, contingency, and temporality; in which value judgments and purportedly objective modeling of the world are synthesized nondeterministically, employing a *principle of insufficient reason* that contrasts sharply with the classical Western notion of rationality based on a *principle of sufficient reason*. Engineering cannot be separated from its social relations, especially from relations of social, economic, and political power in their many guises, or from engineering's transformative effect on human experience as well as, through its participation in technological action, on the physical and social worlds discovered in human experience.

The autonomous character of engineering finds expression in the distinctive intellectual problems that it poses. On the basis of the preceding discussion, three types of problems can be identified: (1) those that challenge traditional epistemology and metaphysics because of the willfulness that is integral to engineering knowledge; (2) those relating to the identification of engineering's *theoria,* or generic perspective on its world; and (3) those associated with the recognition that while engineering has a *theoria,* analogous to but different from that of the physical sciences, unlike science, engineering is quintessentially a *praxis,* a knowing inseparable from moral action. Let me now develop each of these points further.

The distinctiveness of engineering knowledge vis-à-vis traditional epistemology and philosophy of science is that it is a form of knowing in which knowledge claims, ostensibly validated by their "successful" implementation in the form of technologies (artifacts, production processes, or techniques), cannot be separated from the willfulness embedded in selective interest, selective action, and the exercise of power. It is a form of knowledge that develops in the context of an explicitly constrained decision-making process characterized by parameters such as Herbert Simon's notions of "satisficing" and "bounded rationality,"[52] by parochial value judgments reflecting the agendas of the institutions within which engineering is practiced and that find expression in the formulation of engineering specifications, in personnel utilization decisions, in issues of control, in the choices of ends and of means acceptable for achieving them, and in determinations of authority and accountability. Engineering knowledge is distinctive for being driven by the search for explicitly nonunique solutions to arbitrarily defined problems.

Engineering thus poses a new set of epistemological problems deriving from a rationality that is different from that of science. The rationality of engineering involves volition, is necessarily uncertain, transient and nonunique, and is explicitly valuational and arbitrary. Engineering also poses a distinctive set of metaphysical problems. The judgment that engineering solutions "work" is a social judgment,[53] so that sociological factors must be brought directly into engineering epistemology and ontology. The fact that engineering practice entails action on a world also implies that the ontology of engineering cannot be bracketed, so to speak. Engineering does not lend itself to a suspension of metaphysics in the manner of a coherence theory of truth: realism of some sort would seem to be intrinsic to engineering.

It is possible to build further on these observations and to begin to identify the distinctive generic world outlook of engineering, what I have been calling its *theoria*. This would be to the *theoria* of the physical sciences as induction is to deduction, as the pragmatic is to the rational, as nondeterministic design is to deterministic law, as the contextual is to the universal, as the pluralistic and contingent is to the unique and necessary, as the open-ended is to the once-and-for-all, as Ciceronian rhetoric is to Platonic philosophy, as the skeptical is to the certain, as the principle of insufficient reason is to the principle of sufficient reason.

The contrasts set up here overlap and are in the main self-explanatory. The worldview of modern science is captured by a cluster of descriptors that includes deductive, rational (in the classic sense of the term), deterministic, lawful (or law postulating), universal, necessary, uniquely true, timeless. Science is Platonic in its intellectualism, in its embrace of the principle of sufficient reason, and in its pursuit of idealized, mathematical, abstractions as the "true" reality from which the physical world can be deduced.

Engineering, by contrast, embraces what I have elsewhere called the principle of insufficient reason, which, as a matter of fact, dominates our daily lives.[54] That is, we continually experience the need to make deliberate decisions that cannot be made the conclusions of deductive arguments. However thoughtfully we try to act, we never possess enough information to permit acting on logically justifiable grounds, where "logically" means deductively. The prominence of the principle of sufficient reason in the Western intellectual tradition contrasts sharply with the principle of insufficient reason, which is taken seriously only by skeptics, probabilists, and pragmatists, that is, by the minority position within mainstream Western philosophy.

The nondeterministic logic of engineering design and the nonuniqueness of engineering problem specifications and solutions mean that engineers must be guided by a principle of insufficient reason. This makes engineering more akin to the rhetorical tradition of the Sophists—and especially of rhetoric as codified by Cicero and revived in the late Middle Ages[55]—than to classical philosophy, whether Platonic or Aristotelian.

Finally, while engineering poses questions of truth and reality analogous to those posed by science, it also draws into their consideration questions of justice, goodness, beauty (if not in the artistic, then in the philosophic sense of judging that the balance of means and ends is harmonious). For, as Paul Goodman argued twenty years ago, the acknowledgment that engineering is a *praxis,* makes engineering a branch of moral philosophy.[56] Because the practice of engineering always takes place in action contexts, and because the form of its characteristic rationality includes a range of synthetic, personal, and social volitional elements, the practice of engineering, unlike that of science, mathematics, and philosophy, is inseparable from questions of *right* action. As a consequence, and in distinct contrast to science, the problems engineering poses for philosophy have a holistic character.[57] The epistemological and methodological questions raised by engineering knowledge cannot be separated from questions of engineering's ontology, because engineering entails *action on a world;* because of this, engineering intrinsically and inescapably raises moral, political, and aesthetic questions that traditionally remain outside the conduct of science.

The purported value neutrality of the technical is an ideologically motivated stratagem. It serves to insulate from criticism the social factors determining technological action at the expense of misrepresenting engineering. It is a projected fantasy of objectivity, an extension of the ancient Greek notion of the *histor* as an absolute umpire,[58] a notion that survives in the longing for expert witnesses who can resolve value-laden controversies impersonally on the basis of value-neutral facts. The creation of technical knowledge, the selective exploitation of technical knowledge, and the social consequences of that exploitation are mutually interrelated. A philosophy of engineering inevitably opens out to philosophical anthropology; moreover, it cannot take the primarily intellectual and disengaged form characteristic of mainstream Western philosophy and science. Philosophy cannot deal with engineering by remaining aloof, nor can it treat engineering in isolation from its social context as it has done with science. Altogether, engineering poses a formidable challenge for philosophy, not unlike the challenge posed by Richard Rorty in his *Philosophy and the Mirror of Nature*, because it is quite clear that philosophers cannot "solve," in any traditional sense of the term, the comprehensive epistemological cum metaphysical cum political, moral, and aesthetic problems that engineering poses.[59] Philosophy can only help to clarify these problems and to promote thoughtful, but not classically rational, responses to them.

NOTES

1. Steven L. Goldman, "Philosophy, Engineering and Western Culture," in *Broad and Narrow Interpretations of Philosophy of Technology*, ed. Paul T. Durbin (Dordrecht: Kluwer, 1990), pp. 125–52.

2. Steven L. Goldman, "The *Techne* of Philosophy and the Philosophy of Technology," in *Research in Philosophy and Technology* (Greenwich, Conn.: JAI, 1984) 7:130–42.

3. Irvin C. Bupp and Jean-Claude Derian, *Light Water Reactors: How the Nuclear Dream Dissolved* (New York: Basic Books, 1978), especially p. 184. See also Lawrence Lidsky, "Safe Nuclear Power," *The New Republic*, December, 1967, pp. 20–23.

4. Bupp and Derian, *Light Water Rectors*, pp. 42–55.

5. Ibid., pp. 15–41. See also Stephen Del Sesto, *Science, Politics and Controversy: Civilian Nuclear Power in the United States, 1946–1974* (Boulder, Colo.: Westview Press, 1979).

6. Del Sesto, *Science, Politics and Controversy*, pp. 98–99; also, Bupp and Derian, *Light Water Reactors*, pp. 53–55. On the recent NRC decision to drop metal fuel research in favor of mixed oxide fuels, see David Senor, "The Oxide vs. Metal Decision" (Paper prepared for the American Nuclear Society, 1987).

7. Joan Lisa Bromberg, *Fusion: Science, Politics and the Invention of a New Energy Source* (Cambridge: MIT Press, 1983).

8. Paul Freiberger and Michael Swaine, *Fire in the Valley* (Berkeley, Calif.: Osborne/McGraw-Hill, 1984). Michael S. Malone, in *The Big Score: The Billion Dollar Story of Silicon Valley* (Garden City, N.Y.: Doubleday, 1985), focuses on the managers behind the Silicon Valley phenomenon; Howard Rheingold, *Tools for Thought: The People and Ideas Behind the Next Computer Revolution* (New York: Simon and Schuster, 1984) focuses on visionary ideas driving computer development.

9. Freiberger and Swaine, *Fire in the Valley*, p. 20.

10. Tracy Kidder, *The Soul of a New Machine* (New York: Avon Books, 1981).

11. John Logsdon, *The Decision to Go to the Moon* (Chicago: University of

Chicago Press, 1970). See also Walter McDougall, . . . *the Heavens and the Earth: A Political History of the Space Age* (New York: Basic Books, 1985).

12. John M. Logsdon, "The Space Shuttle Program, A Policy Failure?" *Science* 232 (30 May 1986): 1099–1105. See also John M. Logsdon and Ray A. Williamson, "U.S. Access to Space," *Scientific American* 260 (March 1989): 34–40; and Thomas H. Johnson, "The Natural History of the Space Shuttle," *Technology in Society* 10 (1988): 417–24. For a discussion of the interaction of politics and technology in the case of NASA's projected space station, see Sylvia D. Fries, "2001 to 1994: Political Environment and the Design of NASA's Space Station System," *Technology and Culture* 29 (July 1988): 568–93.

13. Nick Kotz, *Wild Blue Yonder: Money, Politics and the B-1 Bomber* (New York: Pantheon, 1988); Eliot Marshall, "Bomber Number One," *Science* 239 (29 January 1988): 452–55.

14. Patrick Tyler, *Running Critical: The Silent War, Rickover and General Dynamics* (New York: Harper and Row, 1986).

15. Ibid., pp. 154–58.

16. For a comparison of the influence of technical knowledge and policy factors in the early days of "Star Wars," see Howard Kwon, "A Report on the Defensive Technologies Study Team of 1983: Examining the Nature of Engineering's Role in the Policy-Making Process" (Paper prepared for the Institute of Electrical and Electronics Engineers, 1987).

17. Lisa Heinz, "Preparing for Science and Engineering Careers" (Paper prepared for the Office of Technology Assessment, 1987); *Federal Scientists and Engineers: 1985* (Washington, D.C.: National Science Foundation, 1986); *Engineering Employment Characteristics* (Washington, D.C.: National Academy of Engineering, 1985).

18. *Report of the Presidential Commission on the Space Shuttle Challenger Accident* (Washington, D.C.: U.S. Government Printing Office, 1987), p. 93.

19. See William L. Bulkeley, "Venturing Out: Computer Engineers Memorialized in Book Seek New Challenges," *Wall Street Journal,* 20 September 1985.

20. Walter Vincenti, "The Davis Wing and the Problem of Airfoil Design: Uncertainty and Growth in Engineering Knowledge," *Technology and Culture* 27 (October 1986): 719n.

21. An insufficiently appreciated influence on managerial decision making is the distinctive style or "culture" of a particular company or governmental agency. The impact on Western Electric engineers, for example, of central AT&T policy decisions is described in Peter Temin (with Louis Galambos), *The Fall of the Bell System* (Cambridge: Cambridge University Press, 1987). Richard Meehan, in *The Atom and the Fault* (Cambridge: MIT Press, 1985), describes how interactions among the "cultures" of the U.S. Geological Survey, the Nuclear Regulatory Commission, Pacific Gas and Electric, and private and academic geology consultants affected the development and regulation of nuclear power in California, especially General Electric's Vallecitos test reactor. Case studies of Exxon and Cadbury, respectively, are in Charles Dellheim, "Business in Time: The Historian and Corporate Culture," *The Public Historian* 8 (Spring 1986): 9–22; and Allen Kaufman and Gordon Walker, "The Strategy-History Connection," *The Public Historian* 8 (Spring 1986): 23–39.

22. David Noble, "Automation Madness," in *Science, Technology and Social Progress,* ed. Steven L. Goldman (Bethlehem, Pa.: Lehigh University Press, 1989).

23. For management's perspective, see H. R. Northrup and M. E. Malin, *Per-*

sonnel Policies for Scientists and Engineers (Philadelphia: University of Pennsylvania Press, 1984), pp. 42–52, p. 305, and Appendix C.

24. See Edwin T. Layton, Jr., *The Revolt of the Engineers: Social Responsibility and the American Engineering Profession*, 2d ed. (Baltimore: Johns Hopkins University Press, 1986), pp. 10–15.

25. Samuel C. Florman, "An Engineer's Comment," *Technology and Culture* 27, no. 4 (October 1986): 681.

26. Layton, *The Revolt of the Engineers*, pp. 10–15.

27. Bruce Sinclair, "Local History and National Culture: Notions on Engineering Professionalism in America," *Technology and Culture* 27 (October 1986): 683–93.

28. Alfred Marshall, *Industry and Trade* (London: Macmillan, 1920).

29. John K. Galbraith, *The New Industrial State* (New York: New American Library, 1968).

30. Langon Winner, oral remarks made at a panel on engineering ethics, Second National Conference on Technological Literacy, Crystal City, Virginia, 1987.

31. See Arthur P. Mollela, "The First Generation: Ussher, Mumford, and Giedion," in *In Context: History and the History of Technology, Essays in Honor of Melvin Kranzberg*, ed. Stephen H. Cutcliffe and Robert C. Post (Bethlehem, Pa.: Lehigh University Press, 1989), pp. 88–105; see also Molella, "Inventing the History of Invention," *American Heritage of Invention and Technology* 4 (Spring/Summer 1988): 23–30.

32. Lewis Mumford, *Technics and Civilization* (New York: Harcourt, 1934); *The Myth of the Machine: I. Technics and Human Development* (New York: Harcourt, 1968); *The Myth of the Machine: II. The Pentagon of Power* (New York: Harcourt, 1970).

33. Jacques Ellul, *The Technological Society*, trans. John Wilkinson (New York: Knopf, 1964). On the influence of Vico, Hegel, and Cassirer, see David Lovekin, *Technique, Discourse and Consciousness: An Introduction to the Philosophy of Jacques Ellul* (Bethlehem, Pa.: Lehigh University Press, 1990).

34. Edward A. Constant, *The Origins of the Turbo-Jet Revolution* (Baltimore: Johns Hopkins University Press, 1980).

35. Brooke Hindle, *Emulation and Invention* (New York: Norton, 1983), p. 128.

36. See the works cited in note 8, above.

37. Thomas Hughes, *Networks of Power* (Baltimore: Johns Hopkins University Press, 1986).

38. Ibid., pp. 42–43.

39. Ibid., p. 462.

40. Ibid.

41. David Noble, *Forces of Production* (New York: Knopf, 1984).

42. David Noble, *America By Design: Science, Technology and the Rise of Corporate Capitalism* (New York: Knopf, 1977).

43. Harry Braverman, *Labor and Monopoly Capital: The Degradation of Work in the Twentieth Century* (New York: Monthly Review Press, 1974); Harley Shaiken, *Work Transformed: Automation and Labor in the Computer Age* (Lexington, Mass.: Lexington Books, 1986).

44. James Beniger, *The Control Revolution: Technological and Economic Origins of the Information Society* (Cambridge: Harvard University Press, 1986).

45. Hughes, *Networks of Power*, pp. 297–305; Thomas K. McCraw, *TVA and the Power Fight 1933–1939* (New York: Lippincott, 1971); and David Lillienthal, *TVA: Democracy on the March* (New York: Harper, 1953).

46. Among many other sources, see John McDermott, "Technology: The Opiate of the Intellectuals," *New York Review of Books,* 31 July 1969; Langdon Winner, *The Whale and the Reactor* (Chicago: University of Chicago Press, 1986); Theodore Rozsak, *Where the Wasteland Ends* (Garden City, N.Y.: Doubleday, 1972); E. F. Schumacher, *Small Is Beautiful: Economics as if People Mattered* (New York: Harper, 1973).

47. For an overview of the Strong Program, see Karin Knorr-Cetina and Michael Mulkay, eds., *Science Observed* (London: SAGE, 1983), pp. 1–18. For an exposition, see Bruno Latour, *Science in Action* (Cambridge: Harvard University Press, 1987). Ron Giere's *Explaining Science* (Chicago: University of Chicago Press, 1988) develops a collateral interpretation in a distinctive way.

48. Trevor J. Pinch and Wiebe E. Bijker, "The Social Construction of Facts and Artefacts: or How the Sociology of Science and the Sociology of Technology Might Benefit Each Other," *Social Studies of Science* 14 (1984): 399–441.

49. Latour, *Science in Action.*

50. Hugh G. J. Aitken, *The Continuous Wave: Technology and American Radio 1900–1932* (Princeton: Princeton University Press, 1985).

51. Goldman, "The *Techne* of Philosophy," p. 125.

52. Herbert A. Simon, *Models of Discovery* (Dordrecht: Reidel, 1977), especially "The Logic of Heuristic Decision-Making," pp. 154–78; see also Simon's *The Sciences of the Artificial* (Cambridge: MIT Press, 1969).

53. Latour, *Science in Action.*

54. Goldman, "The *Techne* of Philosophy," pp. 136–42.

55. David Marsh, *The Quattrocento Dialogue: A Classical Tradition and Humanist Innovation* (Cambridge: Harvard University Press, 1980), especially pp. 1–37.

56. Paul Goodman, *The New Reformation: Notes of a Neolithic Conservative* (New York: Random House, 1970).

57. Goldman, "The *Techne* of Philosophy," p. 139–40.

58. Gerald A. Press, *The Development of the Idea of History in Antiquity* (Montreal: McGill-Queens University Press, 1982), p. 18.

59. Richard Rorty, *Philosophy and the Mirror of Nature* (Princeton: Princeton University Press, 1979), pp. 357–94.

Idealizations and the Reliability of Dimensional Analysis

RONALD LAYMON

INTRODUCTION

Science and engineering both require the actual production of numbers. Theories cannot be tested without theoretically generated targets. The items and processes that are of interest to modern engineering cannot be built or controlled without calculated estimates of performance or behavior. But real computability in science and engineering requires the use of idealizations and approximations. Because idealizations are, strictly speaking, false, and the use of approximations may introduce falsity into a calculation, inferences based on actual calculations will virtually always be unsound. For the scientist, such unsoundness means that theories will be protected from a *modus tollens* disconfirmation. The failed predictions can always be blamed on the false idealizations or the approximations that have been used in conjunction with the theory to generate the prediction.[1] For the engineer, the problem is that of justifying the practical reliability of a calculation that is known from the onset to be untrue. If the calculation proves adequate in some domain of application, the problem becomes one of projecting this success to new and untried areas. Obviously, a great many different types of argument get used in attempts to justify such projectability.[2] A basic rule of thumb, unfortunately not totally reliable, is this: if calculation or analysis T_1 is less idealized and more realistic than T_2, then T_1 can be expected to more reliably project to new cases than T_2. This means that engineers by and large attempt to maximize the realism of their calculations, within of course the constraint of computational cost. Dimensional analysis is from this point of view a godsend precisely because it serves as a *surrogate* for actual derivations. Hence, more realistic analyses can be used than would otherwise be possible. But this something-for-nothing aspect of dimensional analysis does generate suspicion. The bulk of

146

this paper deals with the challenging and original criticisms of dimensional analysis made by T. Ehrenfest-Afanassjewa and N. R. Campbell.[3] I shall show that these criticisms, while substantial, are not decisive because, somewhat ironically, they ignore the use and influence of idealizations and approximations. Exactly how this goes and what its consequences are for the justification of dimensional analysis make up what I hope is the positive contribution of this paper.

Since many readers will not be familiar with the mathematical specifics and applications of dimensional analysis, I have included a primer on its mathematical aspects as well as a brief description of its application to fluid motion in pipes. Since my aims here are primarily introductory I have kept the mathematics simple and informal. The primer should nevertheless provide an adequate basis for the rest of the paper.

A PRIMER ON DIMENSIONAL ANALYSIS

By way of introduction I shall consider a standard textbook example, namely, the determination of the period of a pendulum. This example will be used to introduce the basic concepts and methods of dimensional analysis and to suggest certain features of engineering practice, at least in so far as that practice involves applying science. Imagine that some practical purpose or other is served by knowing the functional dependencies of the period of a pendulum.[4] We could approach the problem of determining this dependency purely experimentally, but the history of science and engineering suggests that as a general method such an approach would be doomed to failure. So we decide to apply Newtonian mechanics to our problem. But in order to begin the application of Newtonian mechanics some idealizations or simplifying assumptions have to be made. Standard introductory analyses usually begin with these idealizations:

- the string is massless and inextensible
- the bob mass is concentrated at a point
- there is no resistance due to the surrounding medium
- there are no hydrostatic effects due to the surrounding medium
- the gravitational field strength is constant

There may be more idealizations implicitly lurking in the standard introductory analysis, but the above list is sufficient to indicate a general feature of applied science; namely, it begins, as does pure science, with idealizations. One shared element of pure and applied science is the successive reduction of these idealizations and their replacement with more realistic counterparts. But while analyses may become less idealized, the use of idealizations and approximations never stops. The significance of this continued use of idealizations for dimensional analysis will become clear in the development of this paper. Given the above idealizations, standard vectorial analysis of the forces involved leads to the well-known differential equation

$$ml \frac{d^2\phi}{dt^2} = -mg \sin \phi$$

where $l =$ the length of the pendulum suspension cord, $m =$ the mass of the pendulum bob, $\phi =$ the angular displacement of the pendulum bob from vertical, and $g =$ the rate of acceleration due to gravitational attraction. The differential equation can be considerably simplified if we utilize the approximation $\phi = \sin \phi$ and treat ϕ as a length. But we must hope that the original equation is insensitive to this approximation. That is, that the solution of the equation as modified by the approximation is itself an approximation of the same order as the angle approximation.[5]

Applying the approximation and canceling the two occurences of mass yields

$$\frac{d^2\phi}{dt^2} = -\frac{g\phi}{l} \tag{1}$$

Now imagine that, as is often the case in real engineering situations, we are unable to solve the resulting differential equation. In such a case dimensional analysis can be used along with experimentation to provide the solution. The first step is to show that the above equation is *dimensionally homogeneous*. That is, it must be shown that the functional dependencies remain the same if different units are used for the fundamental or basic measurement units employed. In the case of (1) the basic quantities are spatial and temporal. Let the original system of units of (1) be transformed according to the following set of scale transformations,

$$t' = Tt, \ l' = Ll, \ g' = Gg,$$

where T, L and G are positive real numbers. So, for example, T of the original units for time are equal to one new or primed unit of time. Since the acceleration due to gravity is already defined in terms of length and time, it is the case that $G = L/T^2$. Applying the transformations to the original units of (1) yields,

$$\frac{d^2(L\phi)}{d(Tt)^2} = - \frac{(L/T^2)g}{L \ l} L\phi$$

Therefore, collecting the transformation coefficients we get,

$$\frac{d^2\phi}{dt^2} = -(T^2/L) \ (L/T^2)(1/L)(L) \frac{g}{l}\phi$$

which, of course, reduces to equation (1). So (1) is dimensionally homogeneous.

More insight can be obtained into the behavior of the differential equation (1) if we do not utilize the relation $G = L/T^2$, and if we do not (as was implicitly done above) identify the transformation coefficients l and ϕ. Consider the following set of transformation coefficients : $\phi' = \Theta\phi$, $l' = Ll$, $g' = Gg$, $t' = Tt$. Applying these transformations to (1) we get,

$$\frac{d^2(\Theta\phi)}{d(Tt)^2} = -\frac{G \ g}{Ll} \Theta\phi$$

In order for (1) to be dimensionally homogeneous it must be the case that

$$\frac{T^2 \quad \Theta}{(L \ /G)\Theta} = 1$$

Since it will become important later, let me point out that this transformation constraint has been derived on the assumption that (1) is dimensionally homogeneous. The next step of the dimensional analysis turns on the fact that since $L = l/l'$, $G = g/g'$, and so on, the transformation constraint just derived entails

$$\frac{t^2 \quad \phi}{l/g\phi} = K$$

where K is some constant, and where t is the time when the displacement is ϕ for a pendulum of length l in gravitational field strength g. If we let t be the period T_0, that is, the time corresponding to maximum displacement, we finally arrive at the desired result, namely that

$$T_0 = K \sqrt{l/g}$$

Strictly speaking, we had to assume that the solution is periodic in the above argument. To be sure, experience shows the phenomenon to be periodic, but this periodicity transfers to equation (1) only on the assumption that the idealizations and approximations used do not eliminate or seriously distort that periodicity. Therefore, experimentation can be used to justify assumptions that solutions exist and have a specific character only insofar as there is evidence to show that the idealizations and approximations used do not distort the analysis in relevant ways. But often such evidence does not exist.

The constant K is undermined by our dimensional analysis. Here is another role for experimentation to play. Namely, to determine K given the hypothesis that the period varies directly as $\sqrt{l/g}$. But the sorts of complication just discussed also affect the determination of such constants because the use of idealizations and approximations can be expected to distort or bias the analysis with respect to the actual phenomenon. I shall return to a consideration of these complications in later sections when I consider the criticisms of Ehrenfest-Afanassjewa and Campbell. But for now let me emphasize that in our example dimensional analysis was used primarily as a surrogate for finding an exact solution to the original differential equation.

The dimensional argumentation above corresponds roughly to that employed by Stokes (and later refined by Helmholtz) with respect to the problem of applying the Navier-Stokes equations to the problem of incompressible fluid flow.[6] To summarize, the first step of such argumentation is to determine the transformation constraints imposed by the assumption that initially considered equations that are taken to govern the phenomenon are dimensionally homogeneous. If these initial equations are differential ones, and if it is assumed that a solution exists, then the transformation constraints can be used to determine the basic form of the relation between quantities of interest. The proposal that this type of argumentation be called *inspectional analysis* will be adopted in the rest of this paper.[7]

While inspectional analysis as understood above is a fairly narrow technical device, it does suggest a more general approach, often referred to as *dimensional analysis per se*. Again, I shall use the pendulum case as an illustration. Assume that we are unable to formulate, let alone solve, the differential equation that governs the (ideal) pendulum. This inability corresponds to the usual case. However, given our previous experience we hypothesize that there exists some dimensionally homogeneous function that determines the period in terms of length *(l)*, pendulum bob mass *(m)*, and the acceleration due to gravity *(g)*. The dimensional coefficients of each of the arguments can be conveniently represented in matrix form.

	T_0	m	l	g
M	0	1	0	0
L	0	0	1	1
T	1	0	0	-2

So, for example, g has the dimensions LT^{-2}, while the pendulum bob mass m will trivially have dimension M. Since Rayleigh was one of the more successful and better-known early proponents of dimensional analysis, and since his version is easily motivated, I shall start with it. We begin by assuming that there exists a function that relates period to m, l, and g, and that it has the form,

$$T_0 = K \, m^x l^y g^z$$

where K is a dimensionless constant. Note that functions of this form will be dimensionally homogeneous.[8] Since the period has dimension T, the following dimensional identity must be satisfied in order to meet the requirement of dimensional homogeneity.

$$T^1 = M^x L^y G^z$$

But since,

$$M^x L^y G^z = M^x L^y (L^1 T^{-2})^z = M^x L^{y+z} T^{-2z}$$

this means that, $0 = x$, $0 = y + z$, and $1 = -2z$. Solving yields $x = 0$, $y = \frac{1}{2}$, $z = -\frac{1}{2}$. Therefore, given our initial assumptions,

$$T_0 = K \, m^0 l^{1/2} g^{-1/2}$$

I shall now consider a somewhat more complicated example that will play an important role in later sections of this paper. In a 1915 article reviewing dimensional analysis, Rayleigh considered the Boussinesq problem of heat transfer from a body to a fluid flowing around the body. If the fluid is assumed to be ideal and incompressible then Rayleigh asserts that

> In these circumstances the total heat *(h)* passing in unit time is a function of the linear dimension of the solid *(α)*, the temperature difference *(θ)*, the stream-velocity *(v)*, the capacity for heat of the fluid per unit volume *(c)*, and the conductivity *(κ)*.[9]

The dimensional matrix for these quantities (where C is temperature, and Q is heat) is

	h	α	v	θ	c	κ
L	0	1	1	0	−3	−1
T	−1	0	−1	0	0	−1
C	0	0	0	1	−1	−1
Q	1	0	0	0	1	1

Assume as an initial hypothesis that

$$h = \alpha^x \theta^y v^z c^u \kappa^v$$

So it follows that

$$1 = u + v$$
$$0 = y - u - v$$
$$0 = x + z - 3u - v$$
$$-1 = -z - v$$

Solving in terms of z one gets $u = z$, $v = 1 - z$, $x = z + 1$, $y = 1$. Therefore,

$$h = \kappa \alpha \theta \left(\frac{\alpha v c}{\kappa}\right)^z$$

Now as Rayleigh observes,

> Sinze z is undetermined any number of terms of this form may be combined, and all that we can conclude is that
> $$h = \kappa \alpha \theta \cdot F(\alpha v c / \kappa)$$
> where F is an arbitrary function of the one variable $\alpha v c / \kappa$. (1915, p. 68)

In essence, Rayleigh is assuming that the functional relations exhibited by Nature can be expanded in a MacLaurin series. A somewhat more general perspective on Rayleigh's method is achieved by first observing that in the pendulum case the equation for the period can be rewritten in terms of a *dimensionless* product of powers of the causally relevant components.[10]

$$m^0 l^{1/2} g^{-1/2} T_0^{-1} = K$$

Similarly, Rayleigh's equation for heat flow can be written,

$$\frac{F(\frac{\alpha v c}{\kappa})}{(\frac{\kappa \alpha \theta}{h})} = 1$$

where $\alpha v c / \kappa$ and $\kappa \alpha \theta / h$ are both dimensionless. Elementary linear algebra is sufficient to show that the number of independent dimensionless products is equal to the number of variables considered minus the rank of the dimensional matrix. So, for example, in the pendulum example, since the rank of the dimensional matrix is three, just one independent dimensionless product can be formed. In the case of Rayleigh's analysis of the Boussinesq problem, the rank of the dimensional matrix is four so that the number of dimensionless products is two,

$$\frac{\kappa \alpha \theta}{h}, \frac{\alpha v c}{\kappa}$$

These examples should suggest the existence of some systematic connection between the number of independent dimensionless products that can be formed and alternative representations of the initially postulated dimensionally homogeneous function. Consider first the case, where f is any dimensionally homogeneous function of n arguments, that is, of the form $q_0 = f(q_1, q_2, ... q_n)$ which has a dimensional matrix of rank n, that is, which essentially involves n fundamental dimensions. The dimensional matrix can be represented thus,

	q_0	q_1	...	q_n
Q_1	b_{01}	b_{11}	...	b_{n1}
Q_2	b_{02}	b_{12}	...	b_{n2}
...				
Q_n	b_{0n}	b_{1n}	...	b_{nn}

It is not hard to show that the function f is equivalent to

$$q_0 = K \, q_1{}^{x_1} \, q_2{}^{x_2} ... q_n{}^{x_n}$$

where (a) the product $q_0{}^{-1} \, q_1{}^{x_1} \, q_2{}^{x_2} \, ... \, q_n{}^{x_n}$ will be the independent dimensionless product (referred to above) for the dimensional matrix, (b) $K = f(1, \, ... \, 1)$, and (c) the x_i are the solutions of the equation set,

$$b_{0i} = b_{1i}x_1 + b_{2i}x_2 + ... + b_{nn}x_i \qquad \text{for } i = 1, ...n$$

In case the dimensional matrix has rank $n-1$, there exists a function F such that the initially postulated dimensionally homogeneous function f can be written in the form

$$q_0 = \Pi_l F \, (\, \Pi_2)$$

where $q_0{}^{-1}\Pi_1$ and Π_2 are the two independent dimensionless products of the dimensional matrix. Similarly, if the rank is $n-2$, then the function can be represented as $q_0 = \Pi_1 F(\Pi_2,\Pi_3)$. These results are special cases of what has become known as the Pi Theorem. (The name results from the use of Π to represent dimensionless products.) Since what has been reviewed so far is sufficient for the purposes of this paper I shall not further develop the intricacies of the Pi Theorem other than to note that it is no longer necessary to make Rayleigh's assumption that the initially postulated function can be expanded in a MacLaurin series. All that needs to be assumed is that the postulated function is dimensionally homogeneous.[11]

THE MOTION OF FLUIDS IN PIPES

Having given the reader some of the mathematical flavor of dimensional analysis, I shall next briefly describe its application to the problem of determining the resistance to fluid motion through pipes. This is one of the earliest and best-known examples of a successful use of dimensional analysis. Rather than dwell on the historical and logical subtleties of Reynolds's classic paper of 1883, I shall simply repeat the dimensional analysis of the problem given by Stanton and Pannell.[12] Assume that fluid resistance per unit area (P) depends solely on the mean fluid velocity (v), pipe diameter (x), viscosity (μ), and density (ρ). Since the rank of the dimensional matrix is three, there will be two dimensionless products $\rho v^2/P$ and $vx\rho/\mu$, where the latter is known as the Reynolds number, henceforth denoted R.

	P	v	x	μ	ρ
M	1	0	0	1	1
M	-1	1	1	-1	-3
T	-2	-1	0	-1	0

Therefore, where f is an undetermined function,

$$P = \rho v^2 f(R)$$

The experimental situation was greatly simplified by dimensional analysis since instead of having to determine the variation of each of the relevant variables against one another, one only had to determine the resistance versus the R number. The results were tremendously gratifying and are summarized in figure 1, which gives in simplified form the classic results of Stanton and Pannell. As reported in their 1914 paper, they conducted experiments with water in pipes that varied in diameter from 0.361 to 2.855 cm., and with air in pipes that varied in diameter from 0.361 to 12.62 cm. Figure 1 also includes the curve calculated from the Navier-Stokes equations in the solvable case of laminar flow.

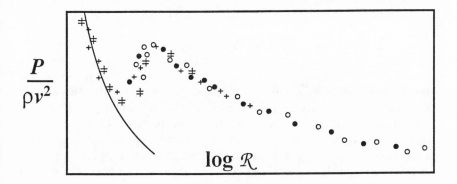

Figure 1

The data indicated that the postulated function f has two branches corresponding to laminar and turbulent motion, and a transition region connecting the two branches.[13] There were, of course, some hitches and infelicities. As later experimentation showed, the exclusive dependency on R broke down for high Mach numbers.[14] There were problems as well in the domain of low Mach numbers since the agreement of different experiments with the same R number occurred only if there was a similarity of pipe surface. From the point of view of dimensional analysis this complication could be viewed as introducing a dimensional constant that was analogous to that used in Hooke's law for springs. Therefore the dependence on the Reynolds number had to be understood as being with respect to similarity classes that were somehow to be defined by the quality of pipe surface.[15] Other breakdowns in the domain of low Mach numbers occurred for capillary tubes.

These failures raise the general question of the projectability of dimensional analyses to untested cases. How reliable are the predictions of functional dependencies? If the only way of determining this reliability were experimental, then dimensional analysis would be of little value to the engineer who desires to construct items or manipulate situations that differ from known examples. Having posed this general question of the reliability of dimensional analysis as an engineering tool, I shall now consider in some detail the criticisms of that reliability made by Campbell and Ehrenfest-Afanassjewa.

CRITICISMS OF DIMENSIONAL ANALYSIS BY CAMPBELL AND EHRENFEST-AFANASSJEWA

As we have seen, dimensional considerations were originally used in lieu of finding solutions to known differential equations and associated boundary conditions. Later developments, in particular the Pi Theorem in its many incarnations, allowed for the application of dimensional analysis in the absence of underlying differential equations. And, as I have tried to demonstrate, the Pi Theorem yields surprisingly strong results for very little investment. This something-for-nothing aspect of dimensional analysis tends to generate suspicion. Its claimed successes are just too good to be true. Those who share this skepticism will be gratified by the energetic criticisms made by N. R. Campbell in his important article of 1924.[16] After noting the dramatic increase in the use of dimensional analysis following Buckingham's 1914 paper on the Pi Theorem,[17] Campbell suggests that (the late) Lord Rayleigh, despite being an early champion of the use of dimensional analysis, would not entirely approve of its newly achieved popularity.

He would probably regret the tendency to expound the principle with great formal generality and considerable mathematical elaboration which, initiated by Buckingham, has led finally to the imposing title of "dimensional analysis"; and he would certainly regret the absurd perversions and erroneous applications of the principle which have been seriously propounded.[18]

This is rather strong language to use regarding the employment of what are rather dry theorems of algebra. Ehrenfest-Afanassjewa, in an elegant and seminal paper of 1926, expressed similar sentiments. Her aim was to free

> so-called "dimensional analysis" from all the mysticism which still clings to it, in spite of the great interest many physicists take in it. In this respect I am encouraged especially by the work of Campbell, in which dimensional analysis is intrinsically nothing less than rejected, and is replaced by what I call the theory of similitude [inspectional analysis]. (1926, p. 258)[19]

My interest is to ascertain what exactly it is about dimensional analysis and its deployment that prompted these strong denunciations. This interest is not merely historical but is prompted by the quality and originality of the underlying criticisms. In Campbell's view the "physical assumptions" that justified the use of dimensional analysis were being obscured by "an undue and entirely unnecessary elaboration" of the "logical" foundations. Hence, Campbell took his aim to be "to draw attention once more to the physical rather than the logical principles upon which 'dimensional analysis' depends." According to Campbell, the application of dimensional analysis requires three physical assumptions.

1. When it is decided that the systems to which the argument is to be applied are physically similar, and have throughout the same no-dimensional magnitudes.
2. When it is decided what are the magnitudes which determine the process discussed.
3. When definite dimensions are assigned to these magnitudes.

Campbell complained that the first and third of these assumptions had not received sufficient consideration.

> In most treatments of the subject, it is the second question which receives most attention: the first is often barely mentioned, and the third is answered simply by the exhibition of a list of all the principal magnitudes together with their dimensions.[20]

Following Campbell's lead, I shall concentrate attention in this paper on the third assumption. That is, I shall ignore, by and large, problems with determining causally relevant factors and shall focus instead on the problem of determining the correct assignment of dimensions to factors already determined to be causally relevant. However, what we shall discover, given this focus, will have ramifications for the division of causally relevant factors into those that can or should be ignored and those which should be included in a dimensional analysis. In order to get some grasp on why Campbell thought the third assumption so important, consider on what basis one assigns the dimensions MLT^{-2} to force. The answer seems obvious: Newton's second law. And this answer seems to be an instance of a general truth: "When we say that a magnitude has certain dimensions we mean that it is involved in some numerical law."[21] But note that force is also "involved" in Newton's law of gravitation. In order to express this law as a dimensionally homogeneous equation between force and the relevant masses and distances, we must introduce a universal constant *with the dimensions* $L^3M^{-1}T^{-2}$. Obviously, this procedure could be reversed and the law of gravitation be taken as primary, that is, as being free of dimensional constants and as defining the units of force. In this case force would be assigned the units M^2R^{-2}. And in fact, this would just be to adopt what is known as the astronomical system of units. If all of this seems to be no more than an illustration of a benignly conventional aspect of science, consider the following quandary that Rayleigh found himself in concerning his treatment of the Boussinesq problem.[22] As shown above, Rayleigh derived by dimensional analysis the equation,

$$h = \kappa\alpha\theta \cdot F(\alpha v c/\kappa)$$

In a brief but important 1915 note, Riabouchinsky noted that if temperature is defined, according to the kinetic theory of gases, as the mean kinetic energy of the gas molecules, then heat and temperature have the same dimensions, namely, those of energy.[23] This means that the relevant fundamental quantities of the Boussinesq problem can be reduced to four.

	h	α	v	θ	c	κ
L	2	1	1	2	-3	-1
T	3	0	-1	-2	0	-1
M	1	0	0	1	0	0

Unfortunately, as is seen in the dimensional matrix, since the rank is three, the number of dimensionless products is now also three.

This means that Riabouchinsky's apparently more sophisticated and better-informed dimensional analysis yields a *less informative* result than Rayleigh's original treatment! That is, the heat transfer varies as an undetermined function of *two* dimensionless products.

$$h = \kappa\alpha\theta \cdot F(v/\kappa\alpha^2, \; c\alpha^3)$$

Rayleigh, in his often quoted response, accepted the paradoxical nature of Riabouchinsky's result, but offered no real resolution.

> The question raised by Dr. Riabouchinsky belongs rather to the logic than to the use of [dimensional analysis], with which I was mainly concerned. It would be well worthy of further discussion. The conclusion that I gave follows on the basis of the usual Fourier equation for the conduction of heat, in which heat and temperature are regarded as *sui generis*. It would indeed be a paradox if further knowledge of the nature of heat afforded by molecular theory put us in a worse position than before in dealing with a particular problem. The solution would seem to be that the Fourier equations embody something as to the nature of heat and temperature which is ignored in the alternative argument of Dr. Riabouchinsky.[24]

From Campbell's perspective, the case illustrated an alarming lack of robustness in dimensional analysis. In the Rayleigh-Riabouchinsky case, a loss of specificity of result was caused by the use of apparently relevant additional information regarding the correct assignment of dimensions to causally relevant factors. On the other hand, as Ehrenfest-Afanassjewa remarked, if truly relevant information is overlooked there will "be projected an illusory definiteness of solution."[25] Campbell and Ehrenfest-Afanassjewa both held the strong position that the only way to prevent such problems is, in Campbell's words, to base dimensional arguments on all and only "the numerical laws or theories by which the process under discussion is determined."[26] According to Campbell, dimensional analysis is "valid" only if

> the laws determining the behavior of the system are completely known, and are actually expressible in the form of differential equations, the integration of which by purely mathematical processes would give a complete solution of the problem.[27]

I shall refer to this thesis as the *fundamental equations thesis*. And I shall assume that by "valid" Campbell means justified or well supported. If this thesis is correct, then it follows that whenever we have a justified use of dimensional analysis we will have available to us "the complete equations" that govern the phenomenon of interest. But if this is so, dimensional analysis is no longer necessary since inspectional analysis can be applied to those equations.

> For if the fundamental differential equation can be stated, all the elaboration of assigning dimensions to the magnitudes in terms of fixed fundamental units is unnecessary, and so is all the mathematical apparatus of Buckingham's Π theorem. The conclusions that it is permissible to draw follow directly from the meaning of similarity, and the identity of no-dimensional magnitudes for similar systems.[28]

It is this implication that Campbell had in mind when he claimed that mathematical elaborations of the Pi Theorem were "entirely unnecessary." I shall refer to this thesis as the *superfluity of dimensional analysis thesis*. I take it that when Campbell says that permissible conclusions follow "from the meaning of similarity, and the identity of no-dimensional magnitudes for similar systems," he means from the decision to apply the same fundamental equations to the phenomena in question and from the identification of the corresponding magnitudes.

THE SUPERFLUITY OF DIMENSIONAL ANALYSIS

Campbell illustrates the superfluity thesis with Stokes's derivation of similarity conditions from the Navier-Stokes equations. Since Ehrenfest-Afanassjewa makes essentially the same point in the context of the simpler pendulum example, I shall return to that case. In my earlier presentation of that example, I noted that if one began with the assumption that the scale transformation for g was LT^{-2}, and that the scale transformation for ϕ was the same as for the pendulum cord length l, then one could show the governing differential equation (1) to be dimensionally homogeneous. Alternatively, one could begin by assuming (1) to be dimensionally homogeneous and then derive the necessary scale restrictions. The latter procedure has the advantage noted by both Campbell and Ehrenfest-Afanassjewa (and illustrated above in the pendulum example) of not requiring the use of additional laws or special assumptions about the correct dimensional decomposition of the quantities employed. In other words, one need not concern oneself about what are fundamental as opposed to derived quantities. This positive feature of inspectional analysis will be further illustrated below in my discussion of projectile motion and Langmuir's law for thermionic current.

There are several disadvantages, however, to inspectional analysis. First and foremost is the fact, as Birkhoff has pointedly observed, "that so few problems . . . have yet been reduced to demonstrably 'well-set' boundary-value problems," that is, where there is a well-behaved combination of governing equations and boundary conditions.[29] Secondly, one must justify the assumption that the fundamental equations used are in fact dimensionally homogeneous. In engineering practice this problem is not of any consequence since most applications of inspectional analysis involve processes that can be taken to be Newtonian. Since there is good reason to believe that Newton's laws are dimensionally homogeneous within the bounds of ordinary engineering applications, it follows that their consequences will be dimensionally homogeneous.[30] The situation for scientific applications is, however, quite different, and assumptions of dimensional homogeneity may be highly suspect. But since the focus of this paper is on engineering applications, I shall not pursue these difficulties of scientific application. The third difficulty with inspectional analysis turns on the fact that those cases where inspectional analysis can be applied inevitably involve the use of idealizations. This means, as noted earlier, that the equations that provide the foundation of the analysis are, strictly speaking, false. One of my aims in this paper is to show how this falsity affects the fundamental equations thesis.

THE FUNDAMENTAL EQUATIONS THESIS

An argument for the truth of the fundamental equations thesis can be developed from a consideration of the simple problem of determining the range of a projectile of mass m, fired at velocity v at angle ϕ. Ignoring air resistance, the relevant quantities and their fundamental constituents are

	x	v	m	g	ϕ
L	1	1	0	1	0
M	0	0	1	0	0
T	0	-1	0	-2	0

Since the rank of this matrix is three, there are two independent dimensionless products, namely, ϕ and xg/v^2. Therefore, if there exists a dimensionally homogeneous function connecting the variables considered, it will be equivalent to one having the form,

$$x = (v^2/g)f(\phi)$$

where f is undetermined. A more specific result can be obtained if one treats horizontal and vertical distance, respectively L_x and L_y, as fundamental and dispenses with the now redundant angular measure ϕ. Since the rank of the resulting dimensional matrix is four, there will be only one dimensionless product, namely, $v_x v_y g^{-1}$.

	x	v_x	v_y	m	g
L_x	1	1	0	0	0
L_y	0	0	1	0	1
M	0	0	0	1	0
T	0	-1	-1	0	-2

Hence, where K is an undetermined constant,

$$x = K(v^2/g)\sin\phi\cos\phi$$

In what follows I shall refer to this second dimensional analysis of projectile motion as the *velocity components analysis* and the first analysis as the *angular analysis*. The question suggested by these alternative dimensional analyses is whether the result of the velocity components analysis is genuinely more informative, or whether the gain in specificity is only illusory. Of course, since there is a standard calculation of the exact result, which is that $x = 2(v^2/g) \sin \phi \cos \phi$, the velocity components analysis can be seen to be more informative. But what reason could be given for thinking it so in the absence of a comparison with the complete calculation? To generalize the question: what sorts of reasons can be given to justify particular forms of dimensional analysis in *actual* cases where *completed calculations are unavailable*? As we have seen, Campbell and Ehrenfest-Afanassjewa held the strong fundamental equations thesis that knowledge of the fundamental equations governing the phenomenon of interest is required. Their claim is not that we need to be able to solve these equations; only that we need to have them available in order to justify the selection of relevant variables and their composition in terms of fundamental quantities. Let us ask then what sorts of reasons, excluding the standard textbook solution, could be given to justify the velocity components analysis of projectile motion. Luce in his discussion of the projectile problem claims,

> [the angular] analysis fails to distinguish direction, even though it is obviously important. The initial velocity has components both in the x and in the y directions, namely, $v_x = v \cos \phi$ and $v_y = v \sin \phi$, the distance traveled has a length component only in the x direction, and gravity has a component only in the y direction. We have ignored all of this information.[31]

At first glance this appraisal seems to refute the fundamental equations thesis since no direct reference has been made to the fundamental equations governing the case. But on closer inspection the situation is not so clear. Why should the extra "information" that Luce points to be relevant for the projectile motion problem? The lesson of the Rayleigh-Riabouchinsky case is *exactly* that apparently relevant additional information may lead us astray. So the *onus* is on critics of Campbell and Ehrenfest-Afanassjewa to explain how the projectile motion case differs from that of the Rayleigh-Riabouchinsky case. In order to give some specific direction to the discussion let us identify the fundamental equations governing the sort of ideal projectile motion being considered. As the reader may recall from his or her undergraduate days, the basic equations for (ideal) projectile motion are

$$\frac{dx}{dt} = v \cos\phi \tag{2}$$

$$\frac{dy}{dt} = v \sin \phi - gt \tag{3}$$

These are the equations that generate the result that $x = 2(v^2/g) \sin \phi \cos \phi$. The thesis to be examined then is whether any reasonable justification of the velocity components analysis will require appeal to these equations. As a first move it could be reasonably claimed that Luce's justification depends on knowing the correct vectorial decomposition of forces, and insofar as knowing the correct decomposition depends on knowing the equations governing the phenomena, to that degree dimensional analysis depends on knowing these equations. If the claim is made in response to this criticism that what justifies the velocity components analysis is that similar velocity decompositions have worked in analogous cases, then I believe we will find that the analogous cases involve the employment of the fundamental equations in question. And if this is so then the case is again made for the fundamental equations thesis. Our problem is to isolate the possible ways of justifying the velocity components analysis *prior* to the empirical testing of its prediction that $x = K(v^2/g) \sin \phi \cos \phi$. There will be at best a fine line between empirical evidence in favor of the dimensional analysis and that in favor of its prediction. We do not want to have to justify the dimensional analysis by having to confirm the prediction. At worst, there is no such distinction to be coherently made. As a concession to critics of the fundamental equations thesis, let us assume that empirical evidence suggests the relevance of v, ϕ, m and g for projectile range. The problem then is whether we opt for the representation of the initial state of the projectile as $\{v, \phi\}$ or $\{v_x, v_y\}$. But if this is the proper way to conceive what is at issue, then there is this puzzle: assuming that experience suggests that m be included in a dimensional analysis, how can that analysis be acceptable if it denies, as both the angular and velocity components analyses do, the relevance of m?

I want to suggest that there is something fundamentally misguided in any attempt to construe the projectile motion case as a counterexemple to the fundamental equations thesis. My basic argument begins with the observation that what is being considered is not real projectile motion but only a sort of counterfactual surrogate. This is why the velocity components analysis is acceptable even though it denies a role for mass. Aristotle may have been mistaken about the influence of mass for counterfactual realms of motion, but he was not mistaken about real motion insofar as mass does play a role. Since the projectile motion being considered is ideal and not real, the only way to gain purchase on the ideal counterfactual situation being considered is to utilize the governing fundamental equations. This is because it is these equations and these equations alone that define the counterfactual situation. There is no independent reality. While all of this may be seen to be good news for the fundamental equations thesis, the bad news is that the confirming case of projectile motion is ideal and not real. And this must raise questions about the truth of the fundamental equations thesis for real as opposed to ideal phenomena. So much for the basic argument. Let us now examine how it plays out in detail.

I shall first develop the sort of mathematical justification that I think is sufficient for demonstrating the superiority of the velocity components analysis. Let me remind the reader that in order to mimic the typical real world situation, I am assuming in what follows that equations (2) and (3) cannot be solved or integrated. Consider the transformation constraints that are derivable from the assumption that (2) and (3) are dimensionally homogeneous for the transformations X, T and G on respectively x, t, and g.

$$\frac{d(Xx)}{d(Tt)} = v \cos \phi$$

$$\frac{d(Yy)}{d(Tt)} = v \sin \phi - GgTt$$

The dimensional homogeneity of (2) immediately yields the constraint that

$$\frac{X}{T} = V$$

However, if we give (2) and (3) the following alternative formulation, and apply the transformations,

$$\frac{d(Xx)}{d(Tt)} = v_x \qquad (4)$$

$$\frac{d(Yy)}{d(Tt)} = v_y - GgTt \qquad (5)$$

we can derive a somewhat stronger constraint, namely that

$$\frac{X}{T} V_x$$

where V_x is the transformation coefficient for v_x. This means that

$$\frac{x}{t} = k v_x = k v \cos \phi$$

where k is an undetermined constant. That a stronger transformation constraint can be so derived is the reason, along with a similarly stronger constraint derivable from (5), that the *corresponding* velocity components (dimensional) analysis also yields a more specific result. That analysis, by postulating a dependence on two types of velocity (v_x and v_y), mimics, so to speak, the stronger of the above two versions of inspectional analysis. If we were to complete an inspectional analysis based on, respectively, equations

(2) and (3) and equations (4) and (5), we would find that the latter
yields, not unexpectedly, the stronger result.[32] So by a comparison
of two forms of inspectional analysis, we come to realize that a
division of the velocity vector into components yields stronger
transformation constraints, and hence a more specific result in
inspectional analysis as well as in corresponding cases of dimensional analysis. But does this justification of the velocity components analysis exhaust the possible justifications as is required by
the fundamental equations thesis?[33] Any proposed alternative justification based on experimental results would have to be highly
suspect because what is being considered is not real projectile
motion but only an extremely idealized, and hence counterfactual,
case, where there is no air resistance and where the only force
acting is a uniform gravitational one. But if all of this is correct, that
is, if it is the ideal nature of the projectile motion case that makes it
supportive of the fundamental equations thesis, then the question is
raised of the correctness of the fundamental equations thesis with
respect to real phenomena. The problem for supporters of the
fundamental equations thesis is to move the argument from considerations of adequacy with respect to ideal phenomena to considerations of adequacy with respect to real phenomena. What the
projectile motion case seems to show is that even if the fundamental equations thesis is true, the justification that is afforded by
reference to the governing fundamental equations is only sufficient
to justify inspectional or corresponding dimensional analyses of
ideal phenomena. Any justification of an inspectional analysis intended to apply to real projectile motion will have to come to grips
with the problem of justifying the use of ideal phenomena as surro-
gates for the actual thing. Unfortunately, the projectile motion case,
as presented, can hardly be expected to offer much guidance with
respect to this problem since it is essentially a textbook example
strongly divorced from its historical origins. In fact, the highly
ahistorical character of my treatment of projectile motion will probably generate suspicion, especially among more historically
minded readers, about the real value and relevance of the preceding
discussion. What is required, therefore, is a more historically sensitive treatment of projectile motion or a move to another example
more closely connected with actual practice. I have chosen the
latter alternative. But before going on to examine that case, I want
to highlight one virtue of the projectile motion case: the division of
the initial velocity vector into vertical and horizontal components
will be retained in more realistic and complete analyses of the
problem. This is to be contrasted with cases where certain aspects

of the analysis are merely artifacts of the idealizations used and disappear once more realistic treatments are given. The relevance of this observation will be seen in our examination of Langmuir's law for thermionic current.[34]

THE LANGMUIR LAW FOR THERMIONIC CURRENT

One of Campbell's aims in his examination of Langmuir's law was to show that "by the method of [inspectional analysis], a correct and new result was obtained [in this case] which was not obtainable (or at least not obtained) by 'dimensional analysis'."[35] Campbell expressed himself somewhat carelessly here since, in fact, Langmuir used neither inspectional nor dimensional analysis in his derivation. Presumably what Campbell wanted to show is that it is in some sense possible to derive the correct form of the Langmuir law by inspectional analysis but not by dimensional analysis. Let us see what this sense might be, for if an appropriate sense of *possible* can be isolated that will make Campbell's claim true, then the case of Langmuir's law supports the contrapositive of the fundamental equations thesis, and hence the thesis itself. This is because what would be shown is that in the absence of a knowledge of the governing equations, it is not possible to justifiably derive the correct form of Langmuir's law by dimensional analysis.

The problem for Langmuir was to determine the thermionic current between a heated cathode and an anode kept at some potential difference. (I shall have more to say about this problem later.) The following dimensional matrix results if we assume that the causally relevant quantities are current density (i), voltage (V), distance (x), mass and charge of the electron (m, e). These quantities and their makeup in terms of fundamental quantities can be represented:[36]

	i	V	x	m	e
M	½	½	0	1	½
L	− ½	½	1	0	³⁄₂
T	−2	−1	0	0	−1

Since the rank is three, two independent dimensionless products can be formed:

$$\frac{e}{Vx}, \frac{V^2}{ix^{3/2}m^{1/2}}$$

So by the Pi Theorem,

$$i = \frac{V^2}{x^{3/2}m^{1/2}} \cdot f(\frac{e}{Vx})$$

Now contrast this approach with that of inspectional analysis. Campbell identifies the "fundamental differential equation" and associated equations as being (where ρ = space charge density, and v = electron velocity)

$$\frac{d^2V}{dx^2} = 4\pi\rho \qquad (6)$$

$$\frac{1}{2} \, mv^2 = Ve \qquad (7)$$

$$i = \rho v \qquad (8)$$

I shall discuss the origin of these equations shortly. But for the moment, note that they can be combined to yield

$$\sqrt{V}\frac{d^2V}{dx^2} = 2\sqrt{2\pi} \cdot i\sqrt{m/e}$$

Assuming this to be dimensionally homogeneous, we can derive the transformation constraint as follows,

$$\sqrt{\Psi V}\frac{d^2(\Psi V)}{d(Xx)^2} = 2\sqrt{2\pi} \cdot Ii \; \sqrt{Mm/Ee}$$

where Ψ, I, M, and E are the transformation coefficients for respectively V, i, m and e. It follows then that

$$\sqrt{\Psi} \, \frac{\Psi}{X^2} \, \sqrt{V}\frac{d^2V}{dx^2} = 2\sqrt{2\pi} \cdot I \, \sqrt{M/E} \cdot i \; \sqrt{Mm/Ee}$$

Therefore,

$$\frac{\Psi^{3/2}}{X^2} = I\sqrt{M/E}$$

So that

$$\frac{V^{3/2}}{x^2} = Ki\sqrt{m/e}$$

As can be seen, this is more specific, though not inconsistent with, the result of dimensional analysis, since

$$i = \frac{1}{K}\frac{V^{3/2}}{x^2}\sqrt{e/m} = \frac{V^2}{x^{3/2}m^{1/2}} \cdot (\frac{e}{Vx})^{-1/2}$$

Campbell's explanation of the greater success of inspectional analysis is this:

> The reason why the dimensional [analysis] . . . leads to an ambiguity is that there has been omitted from it an essential assumption, namely that the stream-lines and lines of force are identical. . . . In this instance it is impossible without examining the equations to realize that there are two lengths involved, parallel and perpendicular to the stream-lines, and that, while similarity in respect of the former is required, similarity in respect of the latter is not.[37]

This makes it appear as if we have a case similar to that of projectile motion, where we failed to distinguish between L_x and L_y. But to categorize the Langmuir case in this fashion is to overlook the fundamentally different ways that initially made idealizations affect the analyses in the two cases. In the projectile motion case, the distinction between vertical and horizontal velocity components will be retained in more realistic analyses, as has already been noted. However, in the Langmuir case, the identity of streamlines and lines of force is, as we shall see, an artifact of the idealization used and is a prime candidate for replacement in more realistic analyses. This means that the greater specificity achieved by the proposed inspectional analysis is illusory and *results from the inclusion of an idealization within the equations on which that analysis is based.*

In order to substantiate and clarify this claim I shall examine in some detail Langmuir's actual calculation. His motivation for making the calculation came from an experimental anomaly dealing with thermionic currents, that is, with currents produced by heated filaments or cathodes. (Such currents form the basis of the ordinary light bulb as well as the vacuum tube.) For sufficiently high voltages (the saturation voltage), the thermionic current is generally found to be independent of the pressure of the intervening gas and increases rapidly with the temperature of the filament. In 1901 Richardson, using a combination (not to be analyzed here) of theoretical and experimental arguments, introduced what has since become known as the "Richardson equation," which relates the emitted thermionic saturation current density i of a heated cathode at absolute temperature T.

$$i = a\sqrt{T} \cdot e^{-b/T}$$

In this equation the terms a and b are experimentally determined constants that vary according to the materials examined. Langmuir, in a series of experiments on tungsten lamps where the voltage difference was less than required for saturation, determined that (1) the Richardson equation fit the experimentally determined data for low temperature; (2) the thermionic current became constant beyond a certain temperature; (3) there was an associated transition curve that connected the region governed by the Richardson equation and that region where the thermionic current was constant. (See figure 2, adapted from Langmuir 1913, p. 451) The problem, as Langmuir saw it, was to determine "the cause of the apparent failure of Richardson's equation at high temperatures."[39]

A series of further experiments revealed that "the factors which governed the value of this new kind of 'temperature' saturation current" included the voltage of the anode, the area of anode, and the distance from anode to cathode.[39] These factors did *not* affect the current in that region governed by the Richardson equation. So the problem was to explain why the initial success of the Richardson equation did not continue into regions of higher temperature. Langmuir hit upon the hypothesis that the failure of that equation was due to a space charge produced by the mutual repulsion of the electrons traveling from cathode to anode. I should observe here that having experimentally isolated the causally relevant factors, Langmuir could have applied dimensional analysis at this stage with good experimental justification. The resulting dimensional matrix, taking the influence of area (A) into account, would have been

	i	V	x	m	e	A
M	½	½	0	1	½	0
L	−½	½	1	0	3/2	2
T	−2	−1	0	0	−1	0

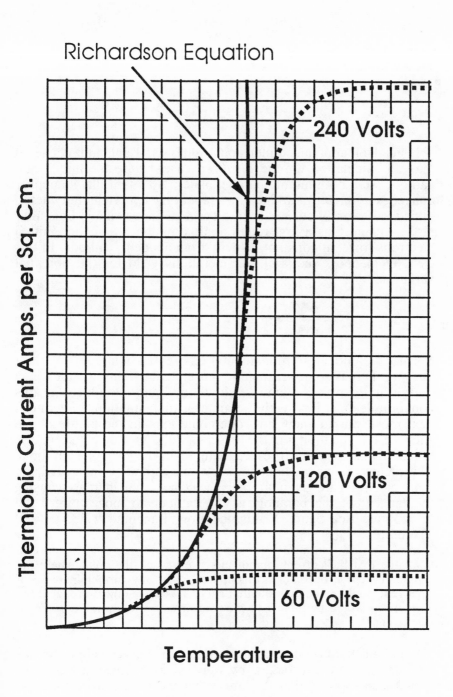

Figure 2

Since the rank is three, there will be three dimensionless products, where

$$i = \frac{V^2}{x^{3/2}m^{1/2}} \cdot f\left(\frac{e}{Vx}, \frac{x^2}{A}\right)$$

Instead of applying dimensional analysis, Langmuir chose to make a calculation for the *idealized* case, where the anode and cathode were both infinite parallel plates. Note that this maneuver concedes *at the onset* taking into account the experimentally determined variation of current with respect to anode size.[40] Langmuir also assumed, "as a first rough approximation," that the emitted electrons travel at constant velocity from anode to cathode. In this case, elementary vectorial considerations yield for the potential variation,

$$\frac{d^2V}{dx^2} = 4\pi\rho \tag{6}$$

This is Campbell's "fundamental differential equation." While I am sure there is some reasonable sense in which this equation may be said to "express" a law, we should not ignore the fact that this equation deals at best with a counterfactual situation. Langmuir's next move was to show how one could explain the failure of the Richardson equation on the basis of (6). As can be seen, the potential distribution expressed by (6) takes the form of a parabola. This means that as the temperature of the cathode is increased and more electrons are given off, the space charge will increase causing the tangent of the parabola at the cathode to move toward the horizontal. (See figure 3, adapted from Langmuir 1913, p. 455) Therefore,

> we see that the current cannot increase beyond the point where the potential curve becomes horizontal at *P*, for any further increase of current would make the potential curve at *P* slope downwards and the electrons would be unable to move against this unfavorable potential gradient.[41]

Langmuir then proceeded to more completely calculate the potential equation. He began by appealing to considerations of Richardson showing that the kinetic energy of an emitted electron is

$$\frac{1}{2}mv^2 = Ve \tag{7}$$

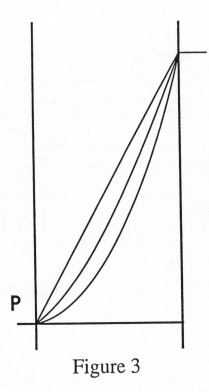

Figure 3

If we combine this with the definition of the current carried by the emitted electrons, equation (8), we have the complete basis of Langmuir's calculation.[42]

$$i = \rho v \qquad (8)$$

Combining (6), (7), and (8) and integrating yields the result,

$$i = \frac{\sqrt{2}}{9\pi} \sqrt{e/m} \, \frac{V^{3/2}}{x^2} \qquad (9)$$

Campbell in effect proposes (or so I interpret him) that had Langmuir been unable to complete the integration, he could have applied inspectional analysis and obtained equation (9), except for the specification of the constant $\sqrt{2}/9\pi$. But Campbell in his discussion of the case does not note that to use equations (6), (7), and (8) as the basis of an inspectional analysis would be to restrict attention to an idealized case and to deliberately ignore the possible variations between electron path and lines of force. Therefore, that "the stream-lines and lines of force are identical" is not a truth about the actual phenomenon but only a truth about its counterfactual surrogate. Hence to claim, as Campbell does, the superiority of his proposed inspectional analysis over the corresponding dimensional analysis is just to accept uncritically the counterfactual surrogate as the real thing! Since Campbell's proposed inspectional analysis is in fact based on the infinite-plane idealization, it is hardly surprising that it leads to a more specific result than that of a dimensional analysis, which takes into account possible effects due to variations in cathode and anode size. In fact, one might argue that the dimensional analysis is superior precisely because it does not utilize the idealization that Langmuir found necessary *for the purpose of completing his calculation.* Langmuir himself was quite careful to emphasize that since the actual thermionic devices considered were not "parallel plane electrodes," his calculation would not "apply rigorously."[43] And when discussing the quality of experimental fit achieved, he says,

> The numerical values obtained from these equations are certainly of the right order of magnitude and agree as well with the experimental results as could be expected when the shape of the electrodes departs so far from those assumed in the calculations.[44]

The reader might wonder at this stage exactly how well is "as well . . . as could be expected." A closely related question is how the experimental divergences from prediction affect Langmuir's proposed explanation of the failure of Richardson's equation. My position on these matters is that justifying the experimental fit and the explanation of the failure of Richardson's equation requires showing that more realistic analyses, if available, could be expected to converge on the actual experimental data. But I shall not explore this type of reasoning here.[45] Instead, I want to develop a variation of Campbell's argument not susceptible to the criticism that he has uncritically confused a counterfactual surrogate for the actual phenomenon. This variation will depend on the assumption that at least in some cases it is preferable to have an equation that is (a) strictly speaking false, (b) precise, and (c) usefully approximate to having one that is by contrast strictly speaking true but vague.[46] In other words, precise but false approximations are sometimes preferable to those that are true but vague. This certainly seems to be frequently the case in engineering situations. If this is so, then the question arises whether idealizations and associated simplifications that sacrifice strict truth for useful inaccuracy can be incorporated into dimensional analysis. Since I cannot hope to answer this question with any sort of completeness here, I shall focus on the more specific question of whether some functional equivalent of Langmuir's parallel-plate idealization could be incorporated into a dimensional analysis of thermionic current.

Restricting attention to ideal infinite planes is precisely to give up any attempt to find similarity laws that deal with variation of anode and cathode size, and to make the distance between cathode and anode the only significant spatial dimension. In order therefore to incorporate this idealization or simplification into a dimensional analysis, we should first isolate the occurrence of *area* in the definition of the quantities in question, track its role in the calculation of dimensionless products, and then somehow discharge it from further consideration. In other words, the idea is to treat area as a fundamental dimension for the purposes of idealizing it away. Adopting such a strategy yields the dimensional matrix,

	i	V	x	m	e	A
M	$\frac{1}{2}$	$\frac{1}{2}$	0	1	$\frac{1}{2}$	0
L	$-\frac{3}{2}$	$\frac{1}{2}$	1	0	$\frac{3}{2}$	0
T	-2	-1	0	0	-1	0
A	-1	0	0	0	0	1

Since the rank of this matrix is four, there will be two dimensionless products, which can be represented,

$$\frac{im^{1/2}x^2}{e^{1/2}V^{3/2}}, \frac{A}{x^2}$$

Therefore,

$$i = \frac{e^{1/2}\ V^{3/2}}{m^{1/2}\ x^2} \cdot f\left(\frac{A}{x^2}\right)$$

Remembering that x is the distance from cathode to anode, we can see that to use the infinite plane idealization is, in essence, just to ignore the possible variation of the ratio A/x^2. Therefore, we can capture the functional equivalent in dimensional analysis of that idealization by simply assuming that the ratio A/x^2 is always unity, that is, that $f(A/x^2) =$ some constant. But if we do this, then we have *exactly* the result that Campbell achieves using inspectional analysis and the idealized differential equation (6).[47] Langmuir's primary motivation for using the infinite plane idealization was to generate an actual calculation of a result even if that result held only of highly counterfactual situations.[48] This suggests that the reason for using particular idealizations will be a function of the specifics of available mathematics and computational capabilities. Therefore, insofar as dimensional analysis will be blind to these specifics, to that degree there will be no motivation to introduce the functional equivalents of useful idealizations into the dimensional analysis. This suggests a modified version of the fundamental equations thesis, namely, that in those cases where the motivation to introduce (ultimately) useful simplifications depends on exact if ideal computations, dimensional analysis will be inferior to inspectional analysis. I have deliberately left open the question of whether there can be motivations for the introduction of simplifications that do not depend on the use of idealizations and fundamental equations. Answering this question must be left for another discussion.[49]

CONCLUDING REMARKS

My principal concern in this paper has been the reliability of dimensional analysis. Is it possible to argue in favor of the reliability of dimensional analyses prior to the empirical testing of their predictions, and if so how? In particular, I have been concerned with the proper assignment of dimensions to quantities assumed to be causally relevant. Ehrenfest-Afanassjewa and Campbell both claimed that a dimensional analysis can be justified only if the equations that "govern" the phenomenon of interest are known. This is what I have called the fundamental equations thesis. In the case of ideal projectile motion, I tried to show that only a comparison of various inspectional analyses is sufficient to justify a choice of dimensional analyses. Therefore, the case provides a confirming instance of the fundamental equations thesis. But if this is correct, it is because the phenomenon being considered is ideal and hence purely counterfactual. Similar considerations apply in the case of Langmuir's law for thermionic current, since what Campbell identifies as the fundamental equation applies only in ideal cases. So in both the projectile motion and Langmuir law cases the justification of the dimensional analysis turns on the phenomenon being ideal and not real. What gets justified is not some application of dimensional analysis to a real phenomenon but only an application to an ideal surrogate. This raises the question of the relation between the actual phenomenon and its ideal surrogate. Therefore, instead of having achieved some definite result regarding our original problem, the justification of dimensional analyses of real phenomena, I have, in typical philosophical fashion, merely exchanged or transformed the problem. There is, or so I hope, some gain to be achieved in this, if only because problems with dimensional analysis can now be seen to be of a piece with a more general problem, namely, the justification of the use of idealizations and approximations. And as noted earlier, because this justification depends in large part on what is required to achieve actual derivations, dimensional analysis, insofar as it is blind to such exigencies, will be without motivation and foundation.

NOTES

I want to express my gratitude to the National Science Foundation (SES-8608167) and Ohio State University for providing research funding for a larger project of which this essay forms a part.

1. See Ronald Laymon, "Idealization and the Testing of Theories by Experimentation," in *Experiment and Observation in Modern Science,* ed. P. Achinstein and O. Hannaway (Cambridge: MIT Press/Bradford Books, 1985), pp. 147–73.

2. The engineer's projectability is analyzed and illustrated in Ronald Laymon, "The Application of Idealized Scientific Theories to Engineering," *Synthese* 81, no. 3 (December 1989): 353–71.

3. T. Ehrenfest-Afanassjewa, "Dimensional Analysis Viewed from the Standpoint of the Theory of Similitudes," *Philosophical Magazine* Ser. 7, 1 (1926): 257–72; and N. R. Campbell, "Dimensional Analysis," *Philosophical Magazine* Ser. 6, 47 (1924): 481–94.

4. In fact, pendulums were used early on to determine variations in gravitational field strength. This use provided a large part of the motivation for the development of increasingly more realistic and complete treatments of pendulum motion. Some of this history is given in Laymon, "Application of Idealized Scientific Theories to Engineering."

5. For an introductory discussion and illustrations of cases where apparently good approximations seriously lead one astray, see C. C. Lin and L. A. Segel, *Mathematics Applied to Deterministic Problems in the Natural Sciences* (New York: Macmillan, 1974), pp. 186–94.

6. G. Stokes, "On the Effect of the Internal Friction of Fluids on the Motion of Pendulums," *Transactions of the Cambridge Philosophical Society* 8 (1850): 8–145. Reprinted with additions in *Mathematical and Physical Papers* (Cambridge: Cambridge University Press, 1901), 3: 1–141; H. Helmholtz, "Ueber ein Theorem, geometrisch ahnliche Bewegungen flussiger Korper betreffend, nebst Anwendung auf das Problem, Luftballons zu lenken," *Wissenschaftliche Abhandlungen* (Leipzig: Johann Ambrosius Barth, 1882), 1: 158–71.

7. The terminological proposal originates with A. E. Ruark, "Inspectional Analysis: A Method Which Supplements Dimensional Analysis," *Journal Elisha Mitchell Science Society* 51 (1935): 127–33. For a more complete description of inspectional analysis see Ehrenfest-Afanassjewa, "Dimensional Analysis"; G. Birkhoff, *Hydrodynamics: A Study in Logic, Fact and Similitude,* rev. ed. (Prince-

ton: Princeton University Press, 1960), pp. 99–103; and T. Y. Na and A. G. Hansen, "Similarity Analysis of Differential Equations by Lie Group," *Journal of the Franklin Institute* 292 (1971): 471–89.

8. The converse, however, is not true. See Birkhoff, *Hydrodynamics*, p. 90.

9. Lord J. W. Strutt Rayleigh, "The Principle of Similitude," *Nature* 95 (1915): 66–68, 644.

10. This is dimensionless because $M^0L^{1/2}G^{-1/2}T^{-1} = M^0L^{1/2}(L^1T^{-2})^{-1/2}T^{-1} = M^0L^0T^0$.

11. The best place to begin study of the development of the Pi Theorem is Birkhoff, *Hydrodynamics*, pp. 87–99. H. L. Langhaar, in *Dimensional Analysis and Theory of Models* (New York: Wiley, 1951), pp. 29–58, gives a good introductory account. For a logically very strict reconstruction see D. Krantz, R. D. Luce, P. Suppes, and A. Tversky, *Foundations of Measurement* (New York: Academic Press, 1971), pp. 454–70. R. L. Causey, *Derived Measurement and the Foundations of Dimensional Analysis*, Technical Report, no. 5: Measurement Theory and Mathematical Models Reports (Eugene: University of Oregon Press, 1967), is also highly recommended for those with a taste for formal reconstructions. P. W. Bridgman, *Dimensional Analysis* (New Haven: Yale University Press, 1931) is still a valuable source, as is E.Buckingham's historically influential article, "On Physically Similar Systems: Illustrations of the Use of Dimensional Equations," *Physical Review* Ser. 1, 4 (1914): 345–76. A brief but useful historical review is given in E. O. Macagno, "Historicocritical Review of Dimensional Analysis," *Journal of The Franklin Institute* 292 (1971): 391–402. E. T. Layton, Jr., "The Dimension Revolution: The New Relations between Theory and Experiment in Engineering in the Age of Michelson," in *The Michelson Era in Ameican Science 1870–1930*, ed. Stanley Goldberg and Roger H. Stuewer (New York: American Institute of Physics, 1988), pp. 23–39, is an excellent overview of engineering applications; and L. I. Sedov, *Similarity and Dimensional Methods in Mechanics*, trans. M. Friedman and ed. and trans. M. Holt (New York: Academic Press, 1959) is the classic engineering textbook. Also very interesting from this point of view of engineering applications is S. J. Kline, *Similitude and Approximation Theory* (New York: McGraw-Hill, 1965).

12. O. Reynolds, "An Experimental Investigation of the Circumstances Which Determine Whether the Motion of Water Shall be Direct or Sinuous and the Law of Resistance in Parallel Channels," in his *Papers on Mechanical and Physical Subjects* (Cambridge: Cambridge University Press, 1900–1903), 3 : 51–105; T. E. Stanton and J. R. Pannell, "Similarity of Motion in Relation to the Surface Friction of Fluids," *Philosophical Transactions of the Royal Society of London* Ser. A, 214 (1914): 513–24.

13. Reynolds's original hypothesis was that this would be the case. For more details on Reynolds's work see J. Allen, "The Life and Work of Osborne Reynolds," in *Osborne Reynolds and Engineering Science Today*, ed. D. M. McDowell and J. D. Jackson (Manchester: Manchester University Press, 1970), pp. 1–82, and M. J. Lighthill, "Turbulence," in the same volume, pp. 83–102.

14. For references and discussion, see Birkhoff, *Hydrodynamics*, pp.107–10.

15. A very precise specification of what is involved in the assumption of similarity classes is given in Causey, *Derived Measurement*, and Krantz, et al., *Foundations of Measurement*, pp. 9–10.

16. See note 3 above.

17. E. Buckingham, "On Physically Similar Systems: Illustrations of the Use of Dimensional Equations," *Physical Review* Ser. 1, 4 (1914): 345–76.

18. Campbell, "Dimensional Analysis," p. 482.

19. Ehrenfest-Afanassjewa, "Dimensional Analysis," p. 258.

20. Campbell, "Dimensional Analysis," p. 482.

21. Ibid.

22. For more details on the astronomical system, see Langhaar, *Dimensional Analysis*, pp. 9–10.

23. D. Riabouchinsky, "The Principle of Similitude," *Nature* 95 (1915): 591.

24. Rayleigh, "The Principle of Similitude," p. 644.

25. Ehrenfest-Afanassjewa, "Dimensional Analysis," p. 266. She also notes that compensating mistakes can lead to a correct result (p. 270).

26. Campbell, "Dimensional Analysis," p. 494.

27. Ibid., p. 486. Ehrenfest-Afanassjewa makes a similar claim in her "Dimensional Analysis," pp. 266–67.

28. Campbell, "Dimensional Analysis," p. 486.

29. Birkhoff, *Hydrodynamics*, pp. 1, 102.

30. For more detail on these matters, see ibid., pp. 100–101.

31. Krantz, et al., *Foundations of Measurement*, p. 475. I have modified Luce's notation slightly.

32. I leave this as an exercise for the reader.

33. Strictly speaking, the fundamental equations thesis allows for other justifications as long as they make essential use of the governing equations.

34. I. Langmuir, "The Effect of Space Charge and Residual Gases on Thermionic Currents in High Vacuum," *Physical Review* Ser. 2, 2 (1913): 450–86.

35. Campbell, "Dimensional Analysis," p. 488.

36. Following Campbell's argument, the dimensional matrix is in terms of the Gaussian system of measurement. See, for example, Langhaar, *Dimensional Analysis*, pp. 130–31, for a description of that system.

37. Campbell, "Dimensional Analysis," pp. 489–90.

38. Langmuir, "Effect of Space Charge," p. 451.

39. Ibid., p. 454.

40. Langmuir's procedure here is predated by a similar calculation given by C. D. Child in "Discharge from Hot CaO," *Physical Review* Ser. 1, 32 (1911): 492–511.

41. Langmuir, "Effect of Space Charge," p. 456.

42. There is also the assumption of the boundary condition $(dV/dx)_o = 0$.

43. Langmuir, "Effect of Space Charge," p. 460.

44. Ibid., p. 461.

45. For discussion of such argumentation, see the following of my essays: those mentioned in notes 1–2, above; "Using Scott Domains to Explicate the Notions of Approximate and Idealized Data," *Philosophy of Science* 54 (1987): 194–221; "Thought Experiments by Stevin, Mach and Gouy" in *Thought Experiments*, ed. Gerald Massey and Tamara Horowitz (Pittsburgh: University of Pittsburgh, forthcoming); and "The Confirmational and Computational Differences between the Social and the Physical Sciences," in *The Israel Colloquium: Studies in History, Philosophy and Sociology of Science*, ed. A. Ullman-Margalit (Dordrecht: Kluwer, forthcoming), vol. 6.

46. An analysis of this division in terms of its confirmational and computational consequences is given in Laymon, "Using Scott Domains."

47. Note that this result is more specific, on the assumption that $x^2/A = K$, than that obtainable from the dimensional analysis which, I suggested earlier, Langmuir could have applied. Campbell gives a related but not identical argument for the

purpose of showing that "it is incorrect to attribute to area (or volume) the dimension L^2 (or L^3)," in "The Dimensions of Area," *Nature* 9 (1922): 110.

48. See the references given in note 45 for discussion of why such initial calculations are important and the role they play in scientific and engineering practice.

49. For an interesting discussion, see Langhaar, *Dimensional Analysis,* pp. 38–41.

Part 3
Value Issues

Real-World Contexts and Types of Responsibility

HANS LENK

Should an employed engineer "blow the whistle," informing the public, in cases where his or her supervisors order secret polluting emissions or a cover-up of activities that pollute the environment? In cases of this sort, nonmoral responsibilities may come into conflict with moral duties. To be precise, loyalty to those supervisors, or to the corporation involved, would seem to conflict with moral responsibilities to protect the public. If it does, *why* does moral responsibility take priority over the responsibility to obey the orders of superiors? Does the corporation—as Peter French has argued—have moral responsibilities?

To cover these and similar cases, it seems imperative to work out a typology of responsibilities. That is what I attempt here, using for the purpose the Analytic Hierarchy Process of Thomas Saaty.[2]

INTRODUCTION

Ethical or moral questions with respect either to technology in general or to the special case of research-and-development institutions are almost always issues of responsibility. Humans are distinctive as moral beings precisely because they can and do assume responsibilities. To be human is to be a moral being. Nonetheless, moral responsibility—which comes into play when other persons or living things are crucially affected, in terms of life and limb or of psychological or social well-being, by one person's or group's activities—is properly distinguished from genuine responsibilities of a nonmoral sort. And the two can come into conflict.

Many scientists, engineers, managers, and politicians think of scientific and technological research as ethically neutral—or value-neutral in a more general fashion. Only the industrialist or politician

183

in charge, they say, should be held responsible for the uses and results of a particular technological project. But this easy way out turns out to be superficial. There is almost always the potential for ambiguity in applying the results of scientific research for the benefit of a particular group—or even of humankind generally; there are a great many interlocking aspects in all applications of science and technology. Moreover, the supposedly clear distinction between basic and applied research is no longer as clear-cut as it once seemed. In the classic example, the physicists involved, during World War II, in the so-called Manhattan Engineering District Project to develop the atomic bomb, quickly learned the intricacies and complexities of responsibility: they lost their influence on the historic application, the actual dropping of the first two atomic bombs. Their immediate reaction to the event was to establish the Union of Concerned Scientists and the Society for Responsibility in Science (the latter pushed mainly by Victor Paschkis). They also set in motion the Pugwash conferences, and, at the one held in 1958, they declared that because scientists have special expertise, they have special responsibilities.

In 1974, the famous Mount Carmel Declaration on Technology and Moral Responsibility spelled out the need for ethical judgment and moral evaluations in a set of stipulations: Technologists, managers, and politicians bear the responsibility for technological developments and applications, and also for their abuses. The declaration goes on to say that these people cannot be excused or escape from political responsibility. Though widely publicized, this manifesto (and others like it) remains fairly vague, avoiding specifics. Is it even clear what "moral responsibility" and "political responsibility" mean? Is everyone equally responsible?

Computer scientist Joseph Weizenbaum,[3] in another declaration, holds everyone to be co-responsible for the state of the world, but this simply will not work. If everyone is responsible for everything, no one is really responsible for anything in particular. One *must* make distinctions.

Responsibility is a function of power, of impact, and of knowledge. The more strategically central is a person's position in terms of power, influence, and responsibility, the greater is that person's degree of responsibility. This notion can be worked out in detail using the hierarchy process of Saaty, which involves the distributive models of graph theory and a presupposed assignment of rights and duties appropriate to different levels. Using this approach, it is possible to spell out a particular individual's responsibility according to how actively he or she participates in a given group activity—including in this the possibility of disturbing or even destroying the system. The person's responsibilities increase in direct proportion to the centrality or level of influence of his or her role.

The basic idea here is that of joint responsibility. Every actor in the system is responsible in proportion to his or her level of practical influence. It is not the case that a single office-holder in a leading position is responsible for everything the system does. Neither, at the opposite extreme, is anyone to be held responsible if he or she can have no influence on the system.

This approach needs one proviso, however—especially with respect to moral responsibility. There can be no vanishing effect, none of the proverbial "large committee irresponsibility," where the larger the committee, the less responsibility each member has. Joint moral responsibility is not a matter of group distribution, trade-offs, or bargaining; it cannot vanish, no matter how great the number of participants. How this can be—avoiding the vanishing effect while distributing responsibility according to level of influence—is something that still needs to be worked out.

Other problems in the distribution of responsibilities also need to be solved; for example, problems of threshold effects, or of cumulative effects—even, in some cases, a synergism of harmful consequences arising from the failures to make an adequate contribution on the part of a great many isolated actors. Think, to continue this last case, of the notorious problem described by Garrett Hardin ("the tragedy of the commons") in terms of pollution coming from ordinary citizens as well as from industry.[4] Again, unpredicted or even unpredictable side effects pose another problem concerning responsibility. How can anyone be held responsible for anything he or she did not—perhaps even could not—anticipate? Humankind today, with all its technologically multiplied capacities to affect the ecosystem, seems to be responsible for more than it can know, foresee, or anticipate. We do, indeed we must, run some technological risks, but we must also proceed cautiously. If we are to encroach on ecosystems, we had better do so in a carefully controlled manner, step by step. Our investigations into susceptibilities and systemic reactions probably should involve approaches such as chaos theory or catastrophe theory—in general, nonlinear, and interdisciplinary approaches.

Furthermore, in order to be able to deal with the complex problems of distributed and conflicting responsibilities, we need to spell out everything known about different kinds of responsibilities and their mutual relationships. Assignment of responsibility to humanity in general, or to "technological man," or to the ultimate decision maker or politician, will not help in solving problems of distribution, resolving conflicts, or assigning responsibilities strategically. General manifestos—for example, the Mount Carmel Declaration and manifestos emanating from the Pugwash conferences—have not generally differentiated types and levels of responsiblity. The various ethics codes of engineering and scientific societies have come the closest to addressing these problems in detail, but even these codes remain too general to address particular issues of assigning responsibility to particular scientists and engineers.

All this is the reason for our effort here. To repeat, it is imperative to work out a typology of responsibilities if we are ever to deal effectively with problems of overlap, conflict, and distribution of responsibilities.

VARIOUS ACTION RESPONSIBILITIES

The concept of responsibility is relational. It designates at least a five-place relation: A person is responsible *for* something (a consequence or result of an act or omission); with respect *to* another person, group, or living being; in a *forum* of judgment or assessment; in view of a *standard;* at a particular *level* of abstraction or perspective. "Being responsible," etymologically, derives from being obliged to *respond,* to give an accounting to someone for some behavior under certain conditions. The objects to whom I may be held (or think myself) accountable may include an organization, corporation, institution, the law, the state, society, humankind or the idea of humanity; it may be a person, the holder of a position such as a parent, or it may be God.

Here I will not go into all the different aspects of all these *relata* (i.e., place-holders in the five-place relation), but it is surely the case that in practical applications all of them need to be taken into consideration. However, I will draw out the implications, perspectives, and levels of the concepts of responsibility especially in terms of the types of responsibility that often conflict with one another. For instance, responsibilities such as contractual obligations may, at a first level, be morally neutral, whereas at some other level they may be morally relevant.

In describing responsibilities, the first and most general level at which to begin is the responsibility for the consequences or results of one's own actions. My technical term for these is *action responsibilities* (see figure 1). Thus, an agent may be held responsible to someone in that forum in which he or she is accountable. For instance, an engineer who designs a dam is responsible—in terms of technical correctness, feasibility, cost, and safety—to supervisors, employers, clients, and the public.

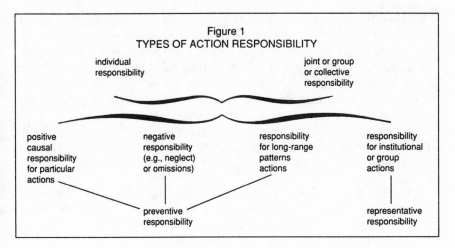

Figure 1
TYPES OF ACTION RESPONSIBILITY

individual responsibility

joint or group or collective responsibility

positive causal responsibility for particular actions

negative responsibility (e.g., neglect) or omissions)

responsibility for long-range patterns actions

responsibility for institutional or group actions

preventive responsibility

representative responsibility

It often happens that accountability questions are raised negatively, pointing out failures in one or more of these respects. In the example of the design engineer, if a dam breaks, it may be a result of miscalculation, careless or negligent workmanship, use of cheap materials, or even cheating. So it is important to emphasize negative (omissions) as well as positive (actions) causal responsibilities. Professionals have a responsibility to the public to maintain high standards in their work and to avoid risks or disasters insofar as that is possible within reasonable cost constraints. This responsibility, though part of causal or action responsibility, is also an instance of role or task responsibility (to be taken up later). *Negative causal responsibility,* in this analytical scheme, is responsibility to avoid careless omissions or neglect.

Another kind of responsibility, taking an active initiative to seek out potential sources of risk or danger, can be termed *preventive responsibility.* For instance, an engineer in charge of quality control must search systematically for things that might go wrong. Here, of course, preventive responsibility is also role responsibility linked formally to a technical job description.

Another distinction needs to be made between action responsibility in the narrow, individualist sense (which could be construed very narrowly, as similar to so-called act utilitarianism) and a general responsibility for patterns of action with *long-range* consequences. For example, parents are responsible for their children not only in terms of individual acts and their consequences but also, in a much more comprehensive perspective, for a great many patterns of actions and omissions.

Corporations and other institutions often act collectively, so there is still another sort of responsibility: *corporate or institutional.* This may coincide, though it is not identical, with the individual responsibility of the person who represents the corporation or institution. *Leadership responsibility,* taken in addressing those outside the institution or representing it in a public forum, is just one example of corporate or institutional responsibility.

Not all types of responsibility involve the individual alone. Joint or group responsibility is involved when members of a group act as a unit. John Ladd has written about collective or group responsibility.[5]

All these distinctions in action responsibility still lie at the most general level of abstraction. We need next to add content to these notions, in terms of role responsibilities and distinctions that have to do with legal as well as moral considerations.

ROLE OR TASK RESPONSIBILITIES

The second level in our typology is that of role and task responsibilities (see figure 2). It is not necessary here to give examples since everyone plays roles and either takes on or is assigned tasks.

In taking on a role—for example, being hired to do a job—a person almost always assumes responsibility for doing the job or playing the role at least at an acceptable level, and sometimes at an optimal level. *Role responsibilities* can be either *formal* or *informal.* They also can be *legally prescribed,* or at least have legal relevance. Furthermore, a person taking on a role may represent his or her corporation or institution and thus may also have *institutional* role responsibilities. A role may involve a *caring responsibility,* for example, a parent's responsibility to care for his or her child's well-being; and institutional roles often carry *legal liability.*

There are also *task responsibilities* associated with the performance of particular tasks—whether these derive from role responsibilities or not. All the same subdistinctions, above, apply here too, including caring responsibilities and legal liability.

Loyalty to persons or institutions need not be connected with particular roles or specific tasks, so another category is needed:

loyalty responsibilities. These, too, involve the various subcategories (above), including caring responsibilities and legal liability.

Finally, at this level, there are also *corporate responsibilities* of institutions, both to insiders and to those outside the institution—for instance, obligations to do certain things for clients, or even for the general public. Subdivisions here include *organizational, moral,* and *legal* responsibilities. And any of these may be a *group responsibility* (in the strict sense discussed above). Those who take on institutional roles may be held *liable,* whether in the strict legal sense, or for moral caring, or for social qualifications—for example, the social control of professional qualifications. In terms of legal liability, whether individual or group, there may be a responsibility to pay a fine, to indemnify persons who have been injured, or to restore an employee to a position from which he or she has been wrongfully removed.

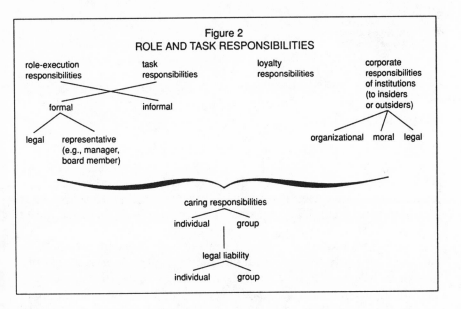

Figure 2
ROLE AND TASK RESPONSIBILITIES

DETAILED MORAL RESPONSIBILITIES

Every particular responsibility cannot be spelled out here, but there are certain *universal moral responsibilities* that constitute the third level in our hierarchy (see figure 3).

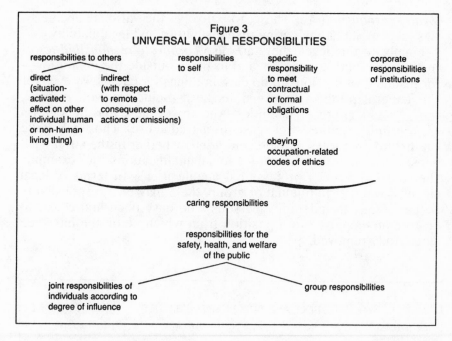

Figure 3
UNIVERSAL MORAL RESPONSIBILITIES

Here again, as with level one, the most obvious responsibility is that of the agent for his or her actions and their consequences in particular situations.

Direct moral responsibility involves other persons or other living beings whose well-being—life and limb, as well as psychological and emotional states—is directly affected by the agent's actions. The so-called moral point of view generally involves other people— or animals—but it can also include responsibilities to *oneself;* that is why a separate category, *responsibilities to self,* is included in the chart.

Remote consequences of an agent's acts or omissions—alone or conjointly with others—introduces the category of *indirect responsibilites.* Neglecting to carry out a safety check or neglectfully stamping an engineering design, say in designing an airplane, can obviously lead to loss of life—as clearly happened with the DC-10 crash of a Turkish Airlines plane outside Paris in 1974. Earlier in 1972, three inspectors at a Long Beach, California, plant had wrongly okayed modifications in the DC-10 cargo door locking system when repairs had not actually been made—with fatal results.[6] A similar case involved falsified certitications of airbrakes by the B. F. Goodrich Company, though this was denied by the company and no loss of life occurred.[7]

More complex questions of indirect *joint responsibility* are raised when the effects are synergistic, or when individual acts only gradually pass a risk threshold—for instance, with respect to pollution or resource depletion in complex ecosystems. (The principle involved here will be addressed in the conclusion.)

It is only recently that *corporate moral responsibilities,* over and above legal responsibilities, have been recognized. The occasions have been failures to improve dangerous conditions: for example, the failures by Convair management in the two DC-10 cases, or by Air New Zealand in a similar case involving a crash on Mount Erebus in Antarctica.[8] This type of responsibility is certainly different from individual moral responsibility. Corporate moral responsibility can coincide or be identical with the *joint moral responsibility,* say, of corporate boards. In my analytical framework, therefore, corporate moral responsibility is put in a category separate from the joint responsibility of group members participating in a group decision or collective action.

Caring responsibility is not limited to roles or specific tasks (level two); it also belongs at this level of concrete applications. The responsibility to care for those dependent on one's actions—whether the dependent is human or not—is not limited to role-based actions but is a universal moral obligation.

Another category involves the specific moral responsibility to meet *contractual and other formal obligations*—from which is derived an obligation to live up to the *ethics code* of one's profession. Again, this is unquestionably a universal moral obligation—provided that meeting the terms of a contract or obeying an ethics code does not conflict with some higher-level moral norm.

Finally, many codes of ethics of engineering and scientific societies include statements about responsibilities for the *safety, health, and welfare* of the public—and some (e.g., that of the IEEE)[9] even say these responsibilities are paramount. Such responsibilities derive from indirect moral responsibilities and from obligations to obey ethics codes—and, as such, are clearly universally binding moral obligations.

CONCLUSION

In sum, responsibility can be broken down into a complex set of levels and types. Moral obligations are just parts of the spectrum, but they deserve some special comments.

Even when moral responsibilities are associated with special roles or actions in particular cases, they remain *universal*. That is, it is not just the specific person involved who is bound by the obligation; the responsibility would apply to anyone in the same role or situation. Morality and moral responsibility are universal. Though individualized distributively, moral responsibility cannot be delegated, substituted, displaced, transferred, or replaced by an individual or an institution. It cannot be diminished, divided up, or dissolved—and it certainly does not vanish, no matter how large the number of people involved. Because it is undiminishable, moral responsibility is also *irreplaceable*.[10]

NOTES

1. Peter A. French, *Collective and Corporate Responsibility* (New York: Columbia University Press, 1984).

2. Thomas L. Saaty, *The Analytic Hierarchy Process* (New York: McGraw-Hill, 1980). Earl MacCormac, currently science adviser to the governor of North Carolina, introduced me to Saaty's approach; in addition, he helped to devise the typology I am presenting here.

3. Joseph Weizenbaum, *Computer Power and Human Reason* (San Francisco: Freeman, 1976).

4. Garrett Hardin, "The Tragedy of the Commons," *Science* 162 (13 December 1968): 1243–1248.

5. John Ladd, "Philosophical Remarks on Professional Responsibility in Organizations," *Applied Philosophy* 1 (1982): 1–13.

6. Paul Eddy, Elaine Potter, and Bruce Page, *Destination Disaster: From the Tri-Motor to the DC-10* (New York: Quadrangle/New York Times, 1976). This case—along with a number of others—is also discussed in Stephen H. Unger, *Controlling Technology: Ethics and the Responsible Engineer* (New York: Holt, Rinehart & Winston, 1982).

7. For references, see Unger, *Controlling Technology*, p. 31.

8. French, "The Principle of Responsive Adjustment: The Crash on Mount Erebus," in *Collective and Corporate Responsibility*, pp. 145–63.

9. See Unger, *Controlling Technology*, pp. 172–79.

10. This paper is an adaptation and abbreviation of my essay "Uber Verantwortungsbegriffe und das Verantwortungsproblem in der Technik," in *Technik und Ethik*, ed. H. Lenk and G. Ropohl (Stuttgart: Reclam, 1987), pp. 112–48; see also my book *Zur Sozialphilosophie der Technik* (Frankfurt: Suhrkamp, 1982).

The Eco-Philosophy Approach to Technological Research

HENRYK SKOLIMOWSKI

THE STRUCTURE OF THINKING IN TECHNOLOGY

We twentieth-century philosophers are positivists—of one variety or another. If we are well educated, it has been drummed into us that we must make sharp distinctions. Thus we must distinguish cognitive from emotive discourse. Very soon we learn that cognitive discourse is all-important, while all other forms of discourse are, well, less important. Very soon also we learn to screen out our emotions, our intuitive hunches, and those deeper intuitive convictions which alone make us unique and articulate human beings. This screening-out process done in the name of the objectivity of research, while innocently proposed as our guide, not infrequently becomes our deity. Thus without our full awareness, let alone our approval, the positivist attitude is thrust upon us. We become its conscious and subconscious heralds.

And such indeed was my personal case. I studied civil engineering at the Warsaw Institute of Technology for five and a half years. Then I taught the subject for the next five years at the same institute. While I was teaching, I pursued my studies in philosophy at Warsaw University, specializing in logic and semantics. Following that, I studied for four years at Oxford University, from which I obtained a doctorate of philosophy in 1964. After those fourteen years of rigorous analytical, objective schooling, I was a herald of Western rationality and positivism.

I started to analyze the phenomenon of technology in the mid-1960s, after I assumed my first university post. Quite predictably, I tried to analyze the phenomenon of technology by concentrating on its cognitive structure. It was a typically positivist exercise, trying to understand the *structure* of the phenomenon (its cognitive structure, that is, and its logical structure—if one can do that) while considering the phenomenon in isolation, as if it were a separate atom, existing in itself. Something similar was going on in the philosophy of science. The luminaries in the field, the Carnaps and the Hempels, tried to understand the cognitive, logical structure of science. I tried to understand the phenomenon of technology by understanding its structure.

However positivist my approach was, I was not an innocent positivist, who simply believed that if one can reconstruct the structure of any phenomenon in some ultimate logical terms, all is well and good. By this time, I had been through Karl Popper's exciting seminar in London, in which a new perspective on the nature of science was revealed and hammered in. The new perspective emphasized that in order to understand the nature of science, one has to understand the nature of its growth. This growth is fluid, dialectical, punctuated with discontinuities. Within the Popperian dialectical perspective, to reconstruct the structure of science in logical terms is an interesting exercise but a bit trivial—for it does not reveal the most important aspect of science, namely, how it grows.

While contemplating what would be the most fruitful way of looking at the phenomenon of technology *in its growth and as distinguished from the phenomenon of science,* I slowly developed the idea that the development of technology is governed by *forms of thinking different* from that of science. The result was my article "The Structure of Thinking in Technology," published originally in 1966.[1] In this early piece I argued that technological progress is the key to the understanding of technology. Without the comprehension of technological progress, there is no comprehension of technology and no sound philosophy of technology. Attempts that aim at reducing technology to applied science fail to perceive that technology is rooted in problems that require solutions.

Although in many instances certain technological advancements can indeed be accounted for in terms of physics or chemistry—in other words, can be seen as based on pure science—it should not be overlooked that the problem that made the advancement necessary was originally not cognitive but technical. With an eye to solving a technical problem, one undertakes inquiries into what is called pure science, with procedures that are extremely selective. Out of a multitude of possible channels of research only very few are chosen. Problems are thus investigated not with an eye to increasing knowledge but with an eye to *solving a technical problem*. If it were not for the sake of solving some specific technological problem, many properties of physical bodies never would have been examined, and many theories incorporated into the body of pure science never would have been formulated.

The growth of technology manifests itself precisely through its ability to produce more and more diversified objects with more and more interesting features, in a more and more efficient way.

It is a peculiarity of technological progress that it provides the means (in addition to producing new objects) for producing "better" objects of the same kind. "Better" may mean many things: more durable, more reliable, more sensitive (if the object's sensitivity is its essential characteristic), faster in performing its function (if its function has to do with speed), or some combination of the above.

In addition to these five criteria, technological progress is achieved through reduced expense or reduced time (or both) in producing a given object.

The criteria of technological progress cannot be replaced or even meaningfully translated into the criteria of scientific progress. And, conversely, the criteria of scientific progress cannot be expressed in terms of the criteria of technological progress. If an enormous technological improvement is made, and at the same time no increase in pure science is accomplished, it will nevertheless mark a step in technological progress. On the other hand, it is of no consequence to pure science whether a given discovery is utilized or not; what is of significance is how much the discovery adds to knowledge, how much it contributes to comprehension of the world.

To summarize, scientific and technological progress are responsible for what science and technology, respectively, attempt to accomplish. Science aims at enlarging knowledge through better and better theories; technology aims at creating new artifacts through means of increasing effectiveness. Thus the aims and the means are different in each case. The kernel of scientific progress can be expressed simply as being in pursuit of knowledge. The answer seems to be less straightforward with regard to technological progress. However, in spite of the diversity of criteria accounting for the advancement of technology, they find a common theme in the measure of effectiveness. Technological progress thus could be described as the pursuit of effectiveness in producing objects.

My second argument in "The Structure of Thinking in Technology" was that specific branches of learning originate and condition specific modes of thinking, develop and adhere to categories through which they can best express their content and by means of which they can further progress. I went on to distinguish various styles of thinking in various branches of technology. Such was my early struggle with the philosophy of technology.

I could have gone on developing this structural-historical approach to technology and its forms of thinking; had I spent ten years with this kind of endeavor, I would probably have convinced myself that this is the best way of studying the phenomenon of technology. Yet I felt, in those early days in the mid-1960s, that it was a limiting approach and that the phenomenon itself was so complex, so all-pervading, so important to the destiny of society and the existential domains of individuals that a much broader approach was needed—a framework of understanding that far transcended the tidy cognitive categories of the structural approach.

Incidentally, after I presented my paper on the structure of thinking in technology at the AAAS meeting in San Francisco in 1965, one of the audience members, Paul Feyerabend, asked me point blank: "How does your philosophy of technology help us to live?" Though stunned by the question, I recognized its validity—although not in the framework of discourse I had outlined for myself at the time.

Around the same time, Patrick Suppes read the paper. When I asked him about his reactions, he said, "Pretty good. But why don't you make it more rigorous?" What he meant was, why not formalize the whole discourse, and express it in neat logical structures as customary among logical empiricists?

After the San Francisco meeting I knew I would take the road Feyerabend suggested. The limitations of logical positivism were too obvious at the time, while the phenomenon of technology was crying for social, existential, and ethical analysis.

A year earlier, in 1964, Jacques Ellul's book *The Technological Society* appeared in English; the French original, entitled *La Technique,* had appeared in 1954. This book presented quite a challenge to my beliefs about technology. I knew I could not remain in the neat boxes of the structural analysis of technology.

A PHILOSOPHY OF NEEDS

I came to believe that to understand the true nature of the phenomenon of technology is not merely to understand the internal development of technology—from one invention to another—one must also understand the peculiar dialogue between technology and society; or more precisely, between the potential of technology and the needs of society.

Technology becomes what society elicits out of its potential

In itself, technology is mute and dumb, like the universe that has not yet developed life. Out of its extraordinary, almost limitless potential, society elicits and articulates certain options. Once certain options are elicited, there is a tendency to keep developing and articulating similar options. Once a certain direction gains momentum, other options are, as it were, suppressed, or at least neglected; some feel that this clearly happened with the nuclear option, which has been developed at the expense of the solar option.

The nature of the dialogue between technology and society is subtle and all-pervading, to the point where society may inhibit the development of some technologies, or of all technology, or it may accelerate or boost their technological development.

Some societies have had much greater technological potential than they deemed it wise to use. Such was the case in ancient Greece during the Hellenistic period, and in China during the fourteenth and fifteenth centuries A.D. Throughout the history of human societies, technology has never been used simply because it is there. It is used if and when society begins to consider it important in the overall social project. The will of society to use or not to use technology is all-important, although it may be indirectly expressed.

That is not how things are seen in the context of the present technological society. Rather, it is assumed, particularly by the protagonists of the technological society, that technology is there to help us to live. On a more sophisticated level, it is argued that technology is here to satisfy human needs. And it is up to humans to determine which needs are important and to then harness technology to this purpose. This argument, particularly the complex notion of *needs,* requires some examination.

In what way do needs determine how technology should develop, and what kind of research should be undertaken? I will attempt here to spell out a philosophy or concept of needs that may be important, not only for guiding research in technology and applied science but also for understanding the direction of technology as we enter the twenty-first century, for understanding the deeper dialogue between technology and society.

To begin with, we should observe that the recognition of genuine needs may sometimes clash with the recognition of human rights. Human rights, once articulated, seem to be a universal category defining us as human beings. In this sense, to deny human rights to a person is to deny his or her humanity.

Yet human rights (or at any rate the appearance of them) can sometimes be violated—for the sake of human dignity and for the sake of maintaining a deeper aspect of our community than can be granted to us by human rights. How can that be? The answer to this question will be given in the context of my theory of *genuine human needs*. I will start with some examples.

Let us suppose a person, say a busy executive, comes home from work and immediately pours himself a martini. We say he is tired, perhaps under stress, and he needs a martini to relax. We recognize this need. Suppose that after finishing the first martini he pours himself a second one. We are still inclined to say he needs it. It is his right to have it. But what do we say if he reaches for the sixth martini and says he needs it? Well, we are less certain to agree that since he *says* he needs it, it is his right to have it, and that it is his *genuine* need. If he reaches for his eleventh martini of the evening, our immediate reaction is to say that it is not good for him, that he had better not drink any more—regardless of his insistence that he *needs* that eleventh martini. We seem to know better—it is no good for him. If we learn that he reaches for eleven martinis every evening, we are assured that "the man has a problem," or more directly, that he is an alcoholic.

Let us now analyze the situation. Who are we to say that the eleventh martini is not good for him, if he insists that it is his need, if he in fact insists on his right to have a twelfth martini? On the surface, we cannot deny him this right, particularly since he demands it explicitly. Yet in our *heart,* and in our *better judgment,* we do deny him this right. Why? Because—we reason—the man does not know what he is doing. He is destroying himself. And he has no business doing that. Twelve martinis in a row is not a genuine need of any normal person.

There are some deeper criteria that speak through us. Those criteria enable us to assert with clarity and certainty—in critical situations—what is good for a person as a human being. On another level, these criteria inform us what are (or can be accepted as) genuine needs—the significant ones, as contrasted with artificial needs, which are trivial at best, destructive at worst.

The overall criterion that works through us is that of *the sanctity of life*. The principle of the sanctity of life, which we share universally, speaks through us when we say to the person on his eleventh martini, "It is not good for you," even if the person insists that it is his right to have the drink, or if he feels he needs it.

To use a more drastic example, let us consider a heroin addict who insists that he needs his heroin fix—regardless of the consequences. Do we recognize this need as valid? Do we recognize his human right to demand what he wants—although the thing he wants may destroy him? Most emphatically, no. What motivates us and gives force to our conviction is *the imperative of the sanctity of life*.

The imperative of the sanctity of life generates its own subimperatives:

- to treat human life with dignity;
- to treat human life with respect;
- to treat human beings as authentic irreducible beings.

We have now obtained a backbone for a philosophy of needs. To begin with, needs are not subjective or relativistic. They are inter-subjective, and rooted in the concept of human nature. Whatever our disagreements about the nature of human nature, the right treatment of human nature is one that respects the imperative of the sanctity of life, which is the foundation of any conception of a life worth living.

Secondly, we now have the basis for the distinction of genuine needs from artificial needs. Genuine needs are those whose satisfaction does not go against the principle of the sanctity of life. Thus, destroying oneself through the use of heroin or through the abuse of alcohol cannot be deemed to satisfy genuine needs. Moreover, in our overall imperative of sanctity, we have the basis for a much sharper delineation of genuine needs from artificial needs. Genuine needs are those whose satisfaction, in the long run, is life-enhancing, or whose satisfaction contributes to a life of quality and meaning. Artificial needs, by contrast, are those whose satisfaction, in the long run, is not life-enhancing. The satisfaction of artificial needs, at best, contributes to the trivialization of human life; at worst, it is life-destroying.

It is a waste of time to quibble about who should determine which life is significant and which is trivial, for it is pretty clear what the difference is, just as it is clear at what point an alcoholic is ruining his life and a drug addict is destroying his. The sanctity of life is a strong principle in most individuals. And it is a beautiful principle indeed. I am not denying that the boundary between genuine needs and artificial ones is difficult to draw in some cases, as the respective merits of the satisfaction of different needs must be weighed carefully—and often on different scales. What I am insisting on is that the distinction can be drawn and that it is a meaningful one.[2] Only after this distinction is drawn can one reexamine the phenomenon of technology and the dialogue between technology and society. Technology is indiscriminate. It produces everything: good, bad, indifferent, life enhancing, trivial, wretchedly unnecessary. But discriminations must be made.

It is my contention that the criterion by which to evaluate a technological process or piece of research in technology is to ask of it: To what degree does it contribute to the satisfaction of genuine needs?

Such a criterion makes it impossible to accept the argument that all technology is good because it satisfies some need (otherwise it would not be here). This argument is unsound and ultimately trivial. So many alleged needs that technology satisfies are trivial needs, superficial needs or what I have termed artificial needs—their satisfaction contributes nothing (or precious little) to the life-enhancing process. So many so-called needs are induced and outrageously manipulated. The fact that people go on satisfying them does not mean that they have inherent grounding in human nature or that they contribute anything to a person's status as an autonomous being.

There are many products and processes, and a great deal of research that goes with them, whose purpose it is to destroy life—such as nerve gases and weapons of mass destruction. From the standpoint of genuine human needs, these technologies are evil; they are not life-enhancing, they cannot contribute to the quality of life. Their only justification is the paranoia of fear, which itself is not a life-enhancing phenomenon but a life-reducing one. The paranoia of fear does not create strengthened security. On the contrary, it weakens security by creating worry and thus draining off mental resources.

There is always a dialogue going on, as I mentioned before, between technology and society. In traditional cultures, society has an upper hand in this dialogue. It is different in American society, which is, not without reason, called the technological society. In American society, technology calls the tune, so much so that its imperatives become society's norms. The dialogue has become lopsided. Deep cultural and social restraints have been eroded in this climate of technological domination.

Advocates claim that technology is the major force of social and existential amelioration. They claim that it satisfies genuine human needs—that *whatever* it satisfies are genuine needs. It is this last claim that needs to be challenged. Because this claim has *not* been questioned, American society has become increasingly trivialized.

Now whoever thinks that he contributes to the well-being of

society, through his research or actual design of technological products, must ask himself at one point or another: What is the net result of my activities? To what aspect of the well-being of society or of individual human life do I contribute, and in what way?

There is one aspect of human needs that needs further reflection. It is claimed that although technology may not be able to satisfy more refined social, cultural, and spiritual needs, it is, at least, an undoubted champion in satisfying basic human needs for food, shelter, and the other material things that constitute a standard of living. I do not deny that these are basic needs, but they are not the only fundamental basic needs.

When I was in India in 1981, I visited Calcutta, and while there I met Mother Teresa, who took me to what is called "the house for the dying." That is what it actually is. Mother Teresa and her sisters, the Missionaries of Charity, pick up from the streets of Calcutta people who are obviously dying—of hunger—and only those. The idea is to give them shelter in the last hour of their dying. But there is more to it than that. Mother Teresa contends that to be unwanted and rejected by everybody is the most terrible thing that can happen to a human being, worse than starvation and death. Thus, one of the fundamental needs, as basic as food or shelter, and perhaps even more so, is *the need to be cared for.* That is what the radiant sisters in the house for the dying attempt to give to the people in the last days and hours of their shattered existence: love and the sense of being cared for. This need—to be cared for as a human being—cannot be measured on any technological scale. Technology has very little to contribute to this need.

Reflection on this need—indeed, the entire discussion of genuine human needs—makes it obvious that technology cannot guide itself. Some deeper principles are required for that purpose. Otherwise, it gropes blindly. Or it arrogantly asserts that what it does is right because technology *always* benefits society.

These conclusions also apply to research in technology and applied science. Left to itself, this research is a ship without a rudder.

TECHNOLOGY CANNOT GUIDE ITSELF; IT NEEDS A COMPREHENSIVE PHILOSOPHY TO GUIDE IT

It is surprising that although the phenomenon of technology has existed for a long time, and has reached an immense scale, there has never been developed any comprehensive technology policy, let alone a comprehensive philosophy of technology. The result is that when things backfire—and they backfire rather often nowadays—people do not know how to respond *rationally*. The buck is often passed from engineers to planners; from planners to politicians; and finally to businessmen. It seems that in this large equation, nobody is specifically responsible for what technology does, and for its consequences in the long run. Technology, as it were, is guiding itself.

My point is that technology can*not* guide itself. It does not have the wisdom, foresight, or social acumen for the task. The free market makes a poor guide indeed, as it leads only to short-term gain. Besides, it is erratic and irrational. Nobody who watched the market crash in 1987 can seriously think otherwise. The rationality of the market is a big myth.

One would still like to think that doing research—whether in technology or in applied science—is a rational affair. But is it? Yes, the rules of the game within any given project are rational. But what strategies guide the choice of projects? There are three imperatives that can guide research in technology and applied science:

1. *An immanent imperative:* Doing what has been done before—"business as usual." The justification for the research lies in the internal rules of the game. Others have done it in this way and we simply follow them. "A million practicing engineers cannot be wrong."

2. *A social imperative:* Doing what is socially desirable. Before undertaking research, the social consequences of the project are assessed. What good will it do for society in the long run?

3. *An ethical imperative:* Doing what is ethically responsible. The ethical imperative goes even beyond the social imperative, as it attempts to ascertain the long-term human consequences of the research. What kind of person is being molded through given research? What kind of human future is the research creating? The ethical imperative is related to genuine human needs. It is unethical to contribute to the creation of technologies that are not life-enhancing in the long run.

These imperatives need to be discussed in some detail. The *immanent imperative* looms so large, and is so universally accepted, that the question may even be raised whether there is any need for any other imperative. After all, researchers know what they are doing. Their research is based on science and logic, and it produces results. That is how things are done. On the face of it, this imperative is convincing and fool-proof. Yet it is flawed and limited.

If researchers are so convinced that they know what they are doing, why are they so often taken by surprise by the consequences of technological developments? If their approach is logical and scientific, why does it so often lead to social calamities? The answer to these questions is that *the internalist approach to technology or the immanent imperative is oblivious to the large-scale consequences of technology.* It is simply beyond the boundaries of the immanent imperative to consider these large-scale consequences. Is it a triumph of transportation technology (even of automotive technology) that all along those excellent freeways of Los Angeles a million cars are moving at the same time? It may be, if we disregard the fact that "the triumph" has created an ecological disaster, a social nightmare of a city, and one of the most alienating of human environments. So where is the triumph of technology? On the planning boards of engineers?

"Wait a moment," a critic might respond, "what you are describing is not technology but a whole social system." Indeed, it *is* a whole social system—of which technology is an inherent part. To disentangle the technological ingredient from the rest of the social system is to play a foolish internalist game.

I know all the neat internalist models. I know how seemingly rational they are, how effortless in their criteria of suboptimization and internal efficiency. But such a concept of rationality is so narrow and self-serving as to be *ir*rational. Is it truly rational to adhere to the rules of the game regardless of who pays the price of future consequences?

Let me put the proposition in more general terms. *We can be rationally stupid. And we can be irrationally wise.* The former case may occur when a researcher develops a new chemical product that dazzles the current market but for which the next generation of babies will pay through their infirmities. The latter case may occur when a group of protestors stages a sit-in at a rare bird sanctuary that is condemned to become a new industrial site. The protestors may know that the cause at hand is unwinnable because a new industrial plant is certainly going to be built there. But they choose to take their stand not because of what is expedient—and therefore seemingly rational—but because of what is right—and therefore wise in the long run. The distinction is fundamental. The failure to recognize it is what is wrong with the instrumental rationality of so much research in technology.

This is the background for discussion of the other two imperatives of research, the *social* and the *ethical*. I will discuss them in the larger context of the socioexistential realities that modern technology has been—sometimes inadvertently, sometimes deliberately—creating.

It is futile to argue about the rationality of technology and its alleged efficiency if one does not take into account the adverse ecological consequences of technological advancements. It is futile to talk about the social benefits of technology if, at the same time, one does not take into account the adverse social consequences of technology. It is futile to praise technology as the savior of humanity if, at the same time, one does not take into account the individual alienation it brings about, if one does not consider how much meaninglessness it creates.

At the very least, one must take into account four large consequences of technology. These are its impacts on ecology, society, human life, and ethical values. Whatever research and development people engage in, there are consequences to the planet that must be considered. The planet Earth is in human custody. If we lose the environment, we lose God, or so maintains the philosopher-theologian Thomas Berry. I would add that if we lose the planet Earth, we lose all our laboratories for research, including our neat rational models. What is at stake here is not spurious moralizing but a clear awareness that insofar as humans affect the well-being of the planet—through research and development and its consequences— humans must take the responsibility for it all. For, who else will do so? It is clear how thoroughly and often adversely technologies can affect the environment. Yet no one wants to take responsibility. Why? The immanent imperative of research encourages this attitude. *Our responsibility is to do the job within a given structure.* But is that the only responsibility researchers have? The argument goes that it is for others to determine the consequences. The folly of this argument is quite apparent. Who are these "others"? Who can determine the consequences if not the people involved in research and development?

Ecology is only one of the four large-scale consequences. Another is society. Most of the arguments I just spelled out vis-à-vis ecology apply to society. What are the social costs of the development of certain technologies? Who will be footing the bill? Is it rational to claim that a given technology is cheap and easily available if the bill to be paid later will be enormous? Let us not fool ourselves that these are "externalities"; they are an inherent aspect of technological progress.

The premise of all technology is that it brings social and human amelioration. Technology advocates are proud to point out some of the lasting and beneficial consequences of technology, giving credit where credit is due. But when it comes to the negative consequences of technology, they look elsewhere to place blame. That kind of attitude is rooted in moral ambiguity and poor reasoning.

So much is said about material amelioration. What about the quality of human life? What about the meaningfulness of human life? Is it rational to look to technology for nothing more than its contributions to the *material* standard of living? Is this not another exercise of moral ambiguity and poor reasoning?

When instrumental values are consistently emphasized, something happens to intrinsic values—those ultimate human values that guide human destinies and the search for meaning and the fulfillment of life. These intrinsic values become themselves "instrumentalized." They are reduced to economic values and accounted for in dollars and cents. And then something happens to the quality and meaning of life: they are shortchanged. The influence of technology has been profound and all-pervading. It has affected the core of human values. It has been emptying lives of inner meaning. It has done so indirectly, by claiming that what *it* provides is most important. Therefore, by comparison, other things are not important.

One must not be naive or foolish enough to argue that values are a matter of people's choices, that people are not compelled to choose the values technology offers them. One cannot argue this while society deliberately sets up structures and choices that compel people to choose what technology offers. Social determinants created by technology are strong indeed. They affect the choices people make, the values they hold, the very core of human life—its meaning, its significance, its worth.

The consequences of technology cannot be bypassed or brushed aside by anybody who claims to do research in a truly rational and conscientious way. Let me express these consequences by means of a diagram (see below).

This diagram expresses the complexity of technology as the phenomenon that interacts with the wholeness of human life. Note that the diagram shows not only the discrete relationships between technology and four specified realms, but it shows one total system. The consequences on ecology affect the social realm. The consequences of technological progress on ecology and society affect human life—and ultimately values. One cannot assess these consequences piecemeal. What is needed is one comprehensive framework in order to see how these things are related. (I am, for

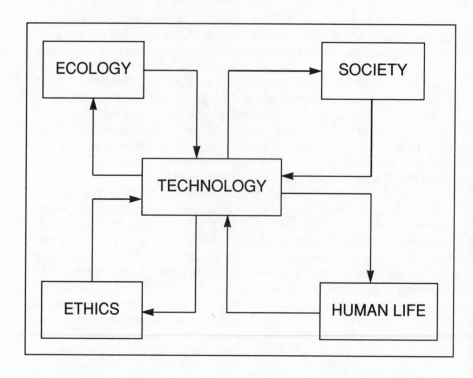

the moment, ignoring other consequences—for example, political consequences—in order not to complicate the picture any further.)

Back in the 1970s, while thinking through those large-scale consequences of technology, I came to the conclusion that what we need is a comprehensive philosophy that would guide and direct technology. Nothing short of this will really be adequate. No tinkering with the internal framework, improving it a little here and there, will do—not even large-scale efforts to humanize technology or tame it. These very expressions, "humanizing technology," or "taming the tiger of technology,"[3] clearly reveal that technology is inhuman. Otherwise, why would one want to humanize it? Why would one want to tame it?

My thinking about a comprehensive philosophy that would effectively guide technology resulted in a large philosophical project that I finally called *Eco-Philosophy*. Simply put, Eco-Philosophy is my philosophical response to the problem of technology as well as to other societal and metaphysical problems that have been plaguing the technological society for quite a while. My Eco-Philosophy was partly created in response to the challenges of Jacques Ellul (mentioned earlier) and E. F. Schumacher—who claimed, in *Small Is Beautiful,* that what we need is not only an alternative economics and an alternative technology but also a metaphysical reconstruction. Our problems are very deep, and the economic-technological framework discussed above cannot do them justice. In due time, I published a book on the subject: *Eco-Philosophy: Designing New Tactics for Living* (1981). The choice of subtitle was quite deliberate: What we need is not another set of philosophical abstractions but a new philosophical framework as a foundation for the right tactics of living.

I will not insist that Eco-Philosophy is the only candidate for a comprehensive philosophy to guide technology.[4] I *do* insist, however, that any comprehensive philosophy make technology accountable to these four realms: ecology, society, human life, ethics.

The ecological consequences of technological progress are not the only ones that should concern us. If these consequences have been thoroughly analyzed and beautifully spelled out, and yet they are still happening and still disastrous, then to know them does not help anyone. We have to find strategies to avoid unintended negative consequences of technology. (The duty to avoid intended negative consequences goes without saying.)

We need an ecology that is healthy and sustainable, a society that is vibrant and cohesive, human life that is pregnant with meaning, human values that do not mock our transcendental and spiritual aspirations. It is against these normative desiderata of the four realms that the development and direction of technology should be assessed.

This, then, is the larger context of the social and ethical imperatives of research. Moreover, the social imperative should not be divorced from the ethical one. They are aspects of each other. This is especially important to realize because the social imperative can be subverted and instrumentalized. When society begins to be dominated by technology and instrumental criteria, our social assessments of technology, after a while, become technological assessments. A lot of activity comes under the label of "technology assessment." Although in principle it is supposed to be social assessment, technology assessment is but lip service paid to societal and ethical concerns. What we witness in fact is mere economic and technological technology assesments. But technology cannot assess itself.[5] Hence, the problem. And hence, social dilemmas and sometimes existential tragedies. Society therefore needs a strong ethical foundation for assessing technology and guiding technological research and development. This brings the discussion back to the principle of the sanctity of life as the basis on which not only to assess genuine needs but also to build the framework of a comprehensive philosophy to guide technology.

The principle of the sanctity of life is thus the cornerstone of a philosophy of technology. If the principle of the sanctity of life is discounted, then, in a sense, nothing follows. The whole technological enterprise does not have much meaning. If the technological enterprise *does* have a meaning and *does* make sense, then it is because it contributes to the well-being of human life. And that implicitly signifies that life is of value, has intrinsic worth. In the ultimate analysis, life must be treated as sacred.

This form of analysis can make many people conditioned by the precepts of objectivity uncomfortable; it seems to suggest some religious underpinning. Actually, it does nothing of the sort. Rather, it reveals that the technological enterprise makes sense only if human life is of intrinsic value. Another way of saying that life is of intrinsic value is to say that it must be not only protected but also revered. At this point, the imperative of the sanctity of life enters as a rational proposition.

Society must not prevent itself from seeing large-scale consequences of technology, including ethical consequences. *Technology is a moral enterprise, and we have every reason to judge it by moral criteria*—and this cannot mean the criteria of technological/instrumental values only. There is actually a deep moral ambiguity, a double-mindedness toward technology, particularly among the protagonists of technology and among big foundations, both private and federal. These foundations are the beneficiaries of science and technology. On the one hand, they attempt to make us see the positive interactions of science and technology with society at large; on the other hand, there is a built-in constraint against seeing that science and technology might have large-scale deleterious consequences for society and human beings. This moral ambiguity pervades academia and a large segment of society. In such a climate, technology is not guided. The tiger of technology is running out of control.

What is required, I would argue, is some in-depth fundamental research in the field. This would be philosophical research that would require ruthless ethical acumen; it would also require open-mindedness on behalf of the sponsors, who would not manipulate scholars to arrive at predetermined conclusions. The sponsors would not, that is, carry on the subtle manipulation that is often present in research on the moral consequences of technology.

The subject of the ethical and human value implications of science and technology is too important to be left in the hands of technicians. Since it is so important, society must be relatively generous and give to people with insight and commitment to human values, a free hand and ample resources. That is the usual procedure for inquiries considered important to society.

There is one more issue to consider. Practitioners and active researchers may get impatient with all these philosophical arguments and may decide the matter is all too complex and difficult. "We should go on with our research without any philosophy," they may protest, and they might add, "What do we need philosophy for if we can do good research without it? We don't need any millstone around our necks." These arguments may *sound* plausible, but *there is no such thing as doing research in our day and age without any philosophy.* Those who claim that they can do their research without philosophy have fallen prey to the philosophy, discussed earlier, of positivism. This is the rationale behind much present-day research: "Give me facts, reason, and the right formula, and I will do the rest." It is the philosphy that implicitly says, "Do not bother me with any philosophy. I have got my philosophy. It is based on the primacy of the physical, the omnipotence of scientific rationality, and the objectivity of results. I do not believe that any normative structure should guide our research." Such are the typical ingredients of positivist philosophy. Unexamined philosophy is not worth having. And so it is with positivism—which is a philosophy for simpletons. As I pointed out in the first section, it is ruefully inadequate for understanding the staggeringly complex issues of our times. To smuggle this philosophy through the back door under the pretense that no philosophy is needed is a naive strategy of the philosophically uneducated. If I appear tough in my statements, I hasten to add that I do not mean to offend anybody. But the issues are tough and complex, and simplistic philosophies for dealing with them simply will not do.

There is no return to innocence. Neither is there a return to the simplicities of positivism. We have created our world of good *and* evil through the unleashing of our technologies. We have to bear the consequences, or amend the situation if we deem it undesirable. "Of all you have done in the past, you eat the fruit, either rotten or ripe," writes the poet T. S. Eliot. Technology will not do the work for us.

SUMMARY AND CONCLUSION

The original premise of *all* technological solutions was the amelioration of the human condition. Thus, the original context of all technology, and of all applied research, was a noble impulse to help the human condition. This original context is the historical mandate that, in public opinion, has never been altered; therefore, it is one to which technologists are accountable.

Research in technology and applied science cannot simply be left to practitioners who would just carry on the work of their predecessors. This is an unenlightened way of doing research. What to do and what not to do is not a matter of little gambits to be confined by the limited frameworks of industrial efficiency; nor is it to be judged by so-called objective criteria of scientific or instrumental rationality—which, with a perfectly rational justification, can lead us to condone the production of weapons of mass destruction.

What to do and what not to do in technology and applied science is a matter of a right vision of humanity and a right philosophy. This philosophy must be comprehensive enough to assess every aspect of a given research project (including its consequences): how it affects individual human beings; the well-being of society; and the health of ecological habitats, including the well-being of all life-forms in the long run. *Only such a philosophy constitutes a framework rational enough to guide our research.*

Thus the framework of this philosophy is normative—even eschatological, since it must evaluate things and relate them to what we deem ultimate destinies of human life on earth. The philosophies that merely justify what exists *because* it exists, and that justify every technological invention as a triumph of progress, in the long run will lead to a head-on collision between what exists and what ought to exist. Love Canal and toxic dumps and nuclear waste sites belong to the reality that is. But their very being is an insult and a threat to human society. In a curious way, they are part of a rational process of doing things effectively within a limited frame of reference. These limited frames of reference cannot be improved if we allow them to operate according to their own terms. They *could* be improved if situated in a larger philosophical framework based on right values.

The recognition of the sanctity of life is a sine qua non of technological responsibility, of technology's responsiveness to human needs. On the basis of the sanctity of life, I have outlined genuine needs, those whose satisfaction is life-enhancing in the long term. The concept of genuine needs, in its turn, should guide and control particular research strategies in the process of implementing various technologies. It is all connected and rational—once we accept the idea of the sanctity of life.

The time of innocence is over. In the unchecked pursuit of

technology, we have signed a Faustian bargain with forces that are beyond our control and sometimes beyond our understanding. To guide ourselves out of this awesome and difficult situation will require more than *mere* rationality—for so often it is this very rationality that is part of our problem. What will be required is foresight, wisdom, vision, and the courage to act in a forthright way. It is possible that the solution itself lies in technology. But it will have to be found and managed by men and women of wisdom.

NOTES

For my contribution to this volume, I was asked to apply my Eco-Philosophy to the special case of the research-and-development community. I did not feel I could do that. Though the leaders of R&D-based corporations, and the politicians with whom they are allied, bear a large part of the responsibility for directing technology, the problem of a philosophy to guide technology is much larger. It is cultural—and, as I say at the end, even eschatological.

1. The article was published in *Technology and Culture* 7 (Summer 1966): 371–83, in a special issue devoted to philosophy of technology. This issue launched the philosophy of technology in the United States as a separate discipline. Incidentally, I was one of the first to use the term "philosophy of technology" to mean an independent philosophical inquiry, separate from philosophy of science.
2. For further discussion of the subject see "A Philosophy of Needs," in my book *Technology and Human Destiny* (Madras: University of Madras, 1983). See also the various writings of Ivan Illich, particularly his *Toward a History of Needs* (New York: Pantheon, 1978).
3. Witold Rybczynski has written a passionate book on this very subject: *Taming the Tiger: The Struggle to Control Technology* (1983). This intelligent book goes far in alerting us to what we can reasonably expect of technology, but, like most other books on the subject, it fails to go deep enough into the philosophical foundations that alone will decide whether technology will get the better of us or whether we will get the better of it.
4. Since I formulated my Eco-Philosophy, other branches of the discipline have come into being, notably Deep Ecology, which, in my opinion, is a bit extreme as a possible guide for technology.
5. For further discussion of the inability of technology to assess itself, see my article, "Technology Assessment as a Critique of a Civilization," in *PSA 1974,* ed. R. Cohen et al. (Dordrecht: Reidel, 1976). In this article I said:

It is quite clear that Technology Assessment has not yet found its identity. However, it is too important a field to be left in the hands of the apologists of Technology. I would therefore like to propose what I call Skolimowski's Laws of Technology Assessment as a set of guidelines for a genuine process of assessing Technology at large. Here are the laws:

(1) No system can adequately assess itself.
(2) The more satisfactory is the assessment from the quantitative point of view, the less valuable it is from the social point of view.
(3) All genuine assessment must terminate in value terms.

(4) The real "expertise" in Technology Assessment is social and moral, not technical (pp. 461–62).

I am not the only one to insist that it is difficult for scientists and technologists to determine the right direction for their research. Albert Einstein once said that because most scientists are totally dependent on society for their support, they are unable to determine the direction of their research.

Part 4
Policy Issues

Judicial Construction of New Scientific Evidence

SHEILA JASANOFF

This essay surveys judicial decisions concerning new scientific evidence as a way of illuminating how scientific research and development (R&D) gains public acceptance in advanced industrial societies. Although this issue holds considerable interest for engineers and scientists, as well as philosophers, neither group has devoted much scholarly attention to interactions between legal decision making and R&D. The case law presented in this essay may well give technical experts some grounds for optimism. At least one example—clinical ecology—suggests that a community of experts acting in concert can influence judicial and public opinion about the reliability of new scientific theories or techniques. Other cases, however, carry a more ambiguous and philosophically more provocative message. The case law on polygraphs and genetic tests, in particular, indicates that judges often draw the boundaries between acceptable and unacceptable evidence differently from technical experts and that disagreements between these two professional communities are not likely to evaporate.

THE SOCIAL CONSTRUCTION OF KNOWLEDGE

A growing body of work in the sociology of science and technology has begun to document and analyze the impact of social and cultural factors on the way scientific knowledge is produced and perceived. This literature provides persuasive evidence that science as we know it cannot be demarcated from its social context and that scientific claims are to a large extent "socially constructed" by actors with different professional and political agendas. Using evidence from areas as diverse as the ethnography of research laboratories[1] and public controversies about risk,[2] the social constructivist school has sought to demonstrate how the social relations of science account for the authority of particular knowledge claims. The constructivist view holds that technology, too, is in some degree a construct of surrounding social paradigms; industrial societies not only produce and use technology, but they invest technology with culturally and historically conditioned meanings, images, myths, and even political force.[3]

Despite the focus on science as a product of social relations, constructivist accounts have paid surprisingly little attention to the role of legal institutions in testing and certifying knowledge claims or otherwise influencing the public acceptance of science and technology.[4] Yet, in the extraordinarily litigious environment of the United States, it is hardly surprising that legal institutions and actors play an active role in the social construction of scientific research and development. Product liability rules provide a simple illustration. This area of law not only determines the commercial prospects for particular products of R&D, such as vaccines,[5] but has created, in the view of some observers, a systematic bias against the development of newer, safer alternatives to hazardous products already on the market.[6]

My aim in this essay is to show that the social construction of scientific knowledge occurs not only in research laboratories and scientific journals but in legal forums such as administrative hearings and courtrooms. Focusing on the latter environment, I argue that the law constructs knowledge, in part, through rules governing the admissibility of evidence. My primary concern is with scientific claims and techniques that the law designates as "novel." This type of evidence presents special challenges to the courts, since it has, by definition, been introduced into legal proceedings before undergoing thorough discussion and review within the scientific community. Cases involving such evidence require judges to function, in effect, as gatekeepers of science. They, and not professional scientists, have to determine as a threshold matter whether the "novel" knowledge claims confronting them are valid and entitled to full respect as "science."

This phenomenon, as I argue below, inevitably draws judges into an activity that sociologists have termed "boundary work."[7] In order to decide questions of admissibility, judges are forced to construct boundaries between what they will and will not accept as science. The occurrence of such boundary-drawing in court proceedings raises critically important questions for both lawyers and scientists. Are the standards of admissibility that courts apply to evidence, especially "novel" evidence, consistent with standards of proof developed by the scientific community? Are the same forms of proof considered authoritative in both legal and scientific settings? What mechanisms exist for resolving conflicts between law and science with respect to the admissibility of evidence? More generally, how does the judicial construction of new scientific claims or techniques affect legal outcomes and the administration of justice?

The essay begins with an account of the legal system's principled attempts to establish the acceptability of scientific and technical evidence. The problems associated with differing legal approaches to this issue are first analyzed in connection with polygraph testing, a technique that remains controversial to this day, though it gave rise to a landmark judicial ruling more than a half-century ago. The essay then examines controversies that have arisen over two more recent types of scientific testimony: the use of DNA fingerprinting in criminal trials and expert claims about ecological illness in toxic tort cases. These examples illustrate some basic divergences between legal and scientific concepts of acceptable evidence. The implications of these findings are explored at the end of the essay.

THE ADMISSIBILITY OF SCIENTIFIC EVIDENCE

A federal court's refusal to admit a primitive lie detector test in 1923 established one of the basic legal tests governing the use of scientific evidence in the courtroom. In *Frye v. United States*,[8] a young black defendant suspected of murdering a white man urged the federal courts in Washington, D.C., either to accept a favorable lie detector test, accompanied by expert testimony as to its reliability, or to let him take the test again in front of the jury. The defense counsel asserted that admitting the test would be consistent with the general rule of evidence allowing experts to testify on matters of specialized experience or knowledge. But a skeptical court concluded that the lie detector had to satisfy a more stringent threshold test of reliability:

Just when a scientific principle or discovery crosses the line between the experimental and demonstrable stages is difficult to define. Somewhere in this twilight zone the evidential force of the principle must be recognized, and while courts will go a long way in admitting expert testimony deduced from a well-recognized scientific principle or discovery, the thing from which the deduction is made must be sufficiently established to have gained general acceptance in the particular field in which it belongs.[9]

In the court's view, the blood pressure detection test relied upon by Frye had not yet gained enough "scientific recognition" to satisfy the test of "general acceptance." Frye was found guilty of second-degree murder and sentenced to life imprisonment but was subsequently exonerated and set free when another man confessed to the crime.[10]

The *Frye* rule, as the general acceptance standard came to be called, has been widely construed as the rule governing the admissibility of new or novel scientific evidence.[11] The opinion grants judges considerable latitude in admitting "expert testimony deduced from a well-recognized scientific principle or discovery." For matters at the dividing line "between the experimental and demonstrable stages," however, *Frye* appears to call for a higher threshold of admissibility. Proponents of such evidence must demonstrate not only that it is reliable and relevant, but that it enjoys near unanimous support from experts in the appropriate fields. According to one authority on trial practice, *Frye* gives technical experts a virtual veto over novel scientific evidence: "dispute or silence among the pundits bars courtroom testimony."[12]

Ever since it was first articulated, however, the *Frye* rule has proved more troublesome to interpret and implement than this analysis implies. Specifically, the injunction that evidence must have support in "the particular field in which it belongs" has acted as a catalyst for judicial disunity. For any type of novel scientific evidence, courts may disagree about what fields of science it "belongs" to, who has the authority to speak for each such field, and how much consensus is needed to establish "general acceptance." Consequently, different jurisdictions have often come to different conclusions about the admissibility of the same evidence. This unsatisfactory state of affairs has gradually undermined judicial and scholarly support for the *Frye* test. By 1985, for example, courts in more than fifteen jurisdictions had rejected *Frye*[13] on the grounds that it leads to inconsistent decisions, promotes excessive judicial deference to scientific opinion, delays the introduction of demonstrably reliable evidence,[14] and forces trial judges to engage in an unrealistic process of "scientific nosecounting."[15]

Courts that regard the general acceptance test under *Frye* as too restrictive have opted in the main for the so-called "relevancy approach" articulated in Rule 702 of the Federal Rules of Evidence.[16] Pursuant to this rule, an expert witness may present any testimony that "will assist the trier of fact to understand the evidence or to determine a fact in issue." Under the relevancy approach, helpfulness to the fact-finder, rather than acceptability to the scientific community, serves as the key to admissibility; expert testimony can be excluded only if its probative value is outweighed by the danger that it might prejudice, confuse, or mislead the jury.[17]

Legal scholars are divided in their views as to whether *Frye* remains a valid test of admissibility even after the enactment of Rule 702. Arguably, the adoption of the Federal Rules of Evidence in 1975 was a sign that Congress intended to repudiate any contrary precedents, such as *Frye*. But although some courts and commentators support this position, others have argued that Congress's failure to say anything about *Frye* at a time when it represented the majority view around the country indicates that the legislature tacitly accepted the ruling.

The relevancy approach of Rule 702 may allow easier entry to some types of scientific evidence than the *Frye* rule. Cases involving the introduction of psychiatric testimony illustrate the salient differences between the two approaches. In *Barefoot v. Estelle,*[18] for example, the Supreme Court held that psychiatric predictions of long-term dangerousness and violent behavior could be presented to the jury, even though the American Psychiatric Association (APA) submitted an *amicus* brief arguing that the accuracy rate for such predictions was no better than one in three.[19] Under the *Frye* standard, the APA submission would very probably have ruled out a finding of "general acceptance" and would therefore have barred the testimony. The Supreme Court, however, refused to exclude psychiatric opinions about future dangerousness, commenting that this would be akin to disinventing the wheel.[20] Instead, the majority resoundingly defended the jury's right to hear the experts on both sides of the dispute:

> [I]t is a fundamental premise of our entire system of criminal jurisprudence that the purpose of the jury is to sort out the true testimony from the false, the important matters from the unimportant matters, and, when called upon to do so, to give greater credence to one party's expert witnesses than another's.[21]

According to this view, the jury has an expansive right to pass on the truth and falsity of expert claims, and professional organizations may not encroach upon this right by imposing on the jury their own independent judgments as to admissibility.

The dissimilarities between the general acceptance and relevancy approaches, however, should not be overdrawn. Both tests betray a fundamentally skeptical attitude toward technical expertise, and both invite judges (and occasionally juries) to engage in boundary work as they differentiate between acceptable and unacceptable expert testimony. The denigration of expertise is perhaps most obvious in the formulation of Rule 702, which treats scientific evidence no differently from any other kind of evidence (i.e., *any* evidence is admissible so long as it is relevant). Under Rule 702, the tasks of deciding what is admissible and what is true belong, respectively, to the judge and to the jury or other legal fact finder. The *Frye* rule on the surface seems more deferential to experts, since it treats the threshold question of admissibility as inherently scientific. But on closer analysis, *Frye*'s deference to the scientific community appears also to flow from a basic mistrust of experts and the authority of science. Without the safeguard of *Frye,* so its advocates argue, the lay fact finder—in most cases the jury—would stand in too much danger of being "taken in by a charlatan in a scientist's smock."[22]

From the standpoint of legal decision making, then, both *Frye* and Rule 702 seem to articulate a fear of being "taken in" by irrelevant or unqualified experts. The remainder of this essay looks at the ways in which the law's skepticism about expert claims has been reconciled with the need to process the increasingly sophisticated proofs offered to courts in this technological age. More particularly, the cases discussed below provide a basis for examining how judges have applied boundary work to decisions about the reliability of scientific evidence and for analyzing the legal and scientific interests that influence the construction of these boundaries.

SEVENTY YEARS OF POLYGRAPHIC EVIDENCE

The polygraph, or lie detector, made its debut in the courtroom in the early 1920s. Invented by William Moulton Marston, a Harvard-trained psychologist and entrepreneur,[23] it was subsequently refined for use in criminal interrogation by John A. Larson of the California police department.[24] The polygraph's role in legal proceedings is based "on the premise that an individual's conscious attempt to deceive engenders various involuntary physiological changes"[25] that can be accurately measured and recorded. Courts, however, have rarely been disposed to regard these instruments as perfect truth-tellers, despite the sweeping claims of the polygraph's inventors. Indeed, it was the D.C. district court's refusal to accept the Marston lie detector that led to the framing of the *Frye* rule.

The judicial response to polygraphs over roughly seventy years provides a useful source of insights into two general questions about the way courts reach qualitative judgments about science and technology. First, how do judges decide who is a trustworthy witness about the reliability of a disputed technique? For example, do court decisions try to incorporate contemporary scientific opinion about the reliability of specific experts or do courts attempt to assess reliability on the basis of independent, judicially constructed criteria? Second, to what extent, if any, have the legal principles discussed above constrained judicial choice as courts weigh conflicting accounts of the reliability of novel scientific evidence?

As noted already, judicial decisions from *Frye* onward have approached the polygraph with greater concern about its deficiencies than faith in its ability to establish the truth. Thus, the *Frye* court itself rejected the Marston test as lacking sufficient scientific "standing" and "recognition." The reasons that the court offered for this conclusion are worth reviewing, because they expose some of the recurring analytical weaknesses that have hampered other courts in their efforts to use the *Frye* rule in a principled fashion.

The bulk of the extraordinarily brief opinion—just nine short paragraphs—was devoted to considering the theoretical foundation for the "systolic blood pressure deception test." As understood by the court,

> the theory seems to be that truth is spontaneous, and comes without conscious effort, while the utterance of a falsehood requires a conscious effort, which is reflected in the blood pressure. The rise thus produced is easily detected and distinguished from the rise produced by mere fear of the examination itself.[26]

The court concluded, with little further analysis, that support for this theory among "physiological and psychological authorities" or in the "discovery, development, and experiments thus far made" was inadequate and did not justify admitting the evidence.

Taking the opinion at face value, it seems clear that the court viewed physiology and psychology as the two fields that are especially relevant to the lie detector's validity. But the opinion neither explained how the court reached this conclusion nor provided any guidance about the criteria that should be used to measure the presence or absence of "general acceptance." By leaving these issues vague and wholly subject to judicial discretion, the *Frye* test set the stage for many decades of inconsistent decision making with respect to the reliability of the polygraph.

Subsequent cases not only failed to establish clearer standards for accepting or rejecting polygraph tests but raised new doubts about whether this technique's admissibility can ever be definitely resolved in accordance with the dictates of *Frye*. In *United States v. DeBetham*,[27] for example, the district court for the Southern District of California concluded that the lie detector was reasonably reliable but still did not meet the *Frye* test. Courts in several other jurisdictions either continued to reject the device as unreliable[28] or decided that it would be impractical and wasteful for courts to try to establish its validity.[29]

If the example of polygraphy is representative, then the passage of time does not necessarily make it any easier for emerging technologies to meet the *Frye* test. More than seventy years after its invention, polygraphy can hardly be regarded as a new technology; nor is there much question that both the instrument and the techniques for administering and interpreting lie detector tests have become more sophisticated during the intervening decades.[30] These developments should in theory have established the reliability and scientific acceptability of polygraphs to the satisfaction of reviewing courts. But the history of litigation over this period shows that the increasing technical refinement of polygraphy has failed to resolve one of the basic dilemmas presented by *Frye*: what kinds of expertise may courts legitimately rely upon when evaluating the admissibility of polygraphic evidence?

To restate this issue in the language of *Frye* itself, courts are as unclear today as they were in 1923 about the "particular field" of science to which polygraphy belongs. This uncertainty may help explain why only a handful of courts have seen fit to reexamine the scientific basis for modern polygraphy in order to determine its admissibility.[31] Before carrying out any such survey, the court must decide as an initial matter where within "science" polygraphy should be located for purposes of technical validation. Yet neither *Frye* nor subsequent cases provide compelling guidelines for identifying the scientific fields "particular" to polygraphy.

One reality that was unrecognized at the time of *Frye*, but that courts now readily accept, is that the examiner's skill and experience bear importantly on the reliability of lie detector tests. Some experienced polygraphists have claimed success rates in the range of 99–100 percent, although critics of the polygraph dismiss such assessments as subjective and of limited probative value.[32] If a court were convinced that the success of polygraphy depends chiefly on the examiner's art and judgment, then it might accept the examiner's credentials as the primary, or even the sole, determinant of admissibility. In practice, however, only a small number of courts have adopted this approach, holding that an experienced polygraph expert's testimony can meaningfully be used to establish the device's reliability.[33]

Sharply diverging from this position, other courts have insisted that polygraphic evidence must be supported not only by the examiner's hands-on knowledge but by a framework of accepted scientific theory. For example, in *United States v. Alexander,* the Court of Appeals for the Eighth Circuit came down firmly on the "polygraphy as science" side of the boundary. In this court's view, the

> polygraph technique is premised upon a complicated interrelationship of psychological stress, a concomitant effect upon the autonomic nervous system and a physiological response. . . . Experts in neurology, psychiatry and physiology may offer needed enlightenment upon the basic premises of polygraphy. Polygraphists often lack extensive training in these specialized sciences. . . .[34]

Expanding upon *Frye,* the *Alexander* court recognized neurology, in addition to psychiatry and physiology, as a scientific field pertinent to polygraphy. At the same time, by emphasizing the complex theoretical underpinnings of polygraphy, the court effectively ruled out mere technical mastery of the instrument as a suitable basis for establishing its validity.

Given the conceptual difficulties that still surround the reliability and acceptability of polygraphic evidence, it is hardly surprising that the law in many jurisdictions has preferred to sidestep these issues altogether. As of 1982 some twenty-five states admitted polygraphic evidence only when both prosecution and defense stipulated in advance that test results should be admitted.[35] Admissibility in these states is founded not on some "objective" criteria of scientific merit announced by experts but simply on the fact that the parties have agreed to abide by the results of testing.[36] For technical experts, this approach to dealing with the problems of polygraphy is full of irony. It shows that the legal system, in its zeal to avoid unnecessary conflict, is prepared to exclude experts altogether from the process of social construction by which the defense and the prosecution come to terms concerning the admissibility of the polygraph.

At present, only two states, Massachusetts and New Mexico,

admit polygraph tests at the request of the defense in criminal cases, even when the prosecution objects. The Massachusetts Supreme Judicial Court held in 1978 that polygraphic evidence was too unreliable for use by the state but could be introduced by defendants provided they agreed to take the stand at trial.[37] In this extreme example of social construction, one party to the legal controversy—the defense—is granted an exclusive right to determine the issue of reliability in keeping with its own interests. More recently, however, attacks on the reliability of polygraphs have led at least one Massachusetts court to conclude that the entire rationale for using polygraph tests in this way should be reconsidered, though this court was unwilling to adopt the stipulation approach as requested by the prosecution.[38]

GENETIC IDENTIFICATION TESTS

The confused state of the law with respect to polygraphy can be contrasted with the relatively smooth passage that the courts have granted thus far to a novel technique known as "DNA fingerprinting." Heralded as the greatest breakthrough in forensic science since fingerprinting,[39] this technique is likely to prove of special utility in identifying the victims and perpetrators of violent crimes and in disputes involving family relationships. In particular, small samples of blood, semen, or hair frequently found at the scene of a rape or murder can be used to identify suspects with virtually no possibility of error. Criminal convictions have already been secured in several U.S. state courts and in Britain on the basis of this technique.

Like the polygraph, DNA fingerprinting was an outgrowth of work done by an individual inventor, Alec Jeffreys. Unlike Marston and the other early polygraphists, however, Jeffreys hit upon his analytical method in the course of doing basic scientific research, in this case, work on certain nucleotide sequences in the human genome. His claims about the technique's reliability accordingly rested from the start on a foundation of theories and methods that have gained wide acceptance in the fields of molecular biology and biotechnology.

The basic discovery underlying the Jeffreys technique for DNA-typing is that the human genome contains certain "minisatellite" regions consisting of repeating "core" sequences of DNA, which vary in length from one individual to another. To isolate this pattern, fragments of DNA obtained through "restriction fragment length polymorphism" (RFLP)[40] are separated by means of gel electrophoresis into bands corresponding to different sizes.[41] These invisible bands are then exposed to radioactively labeled "probes" (short sequences of single-stranded DNA), which bond to specific DNA sequences and create a bar pattern that becomes visible when placed against X-ray film. This pattern is the genetic "fingerprint" that uniquely distinguishes one individual from another.

While enthusiasts for this identification technique stress its infallibility, some influential law enforcement officials have cautioned against adopting it precipitately.[42] Their concern can best be understood if we look back at the checkered history of judicial responses to genetic marker tests before the advent of molecular biological techniques such as RFLP and DNA probes: specifically, the analysis of genetic markers in blood and semen through electrophoresis.

Blood Typing and the Frye Standard

Since the discovery of the ABO system for classifying blood groups at the turn of the century, much progress has been made in analyzing and typing the proteins and enzymes contained in human blood. The polymorphisms, or molecular variations, in these blood factors prompted scientists to theorize that no two individuals have exactly the same constituents in their blood.[43] As in the case of DNA fingerprinting, electrophoresis was the technique by which different blood factors were separated into bands; the claimed uniqueness of each resulting band pattern formed the basis for identifying the donor. The attempt to introduce such evidence into legal proceedings, however, led to confusing decisions that again underscore the difficulty of reconciling legal and scientific standards of admissibility in emerging areas of science and technology.

Courts have offered two major reasons for hesitating to accept blood typing by genetic markers. The first and less controversial has to do with physical factors—such as drying, aging, or contamination of test samples—that may interfere with the reliability of the method.[44] Evidence of such deterioration, and the consequent risk of false positive identifications, might provide reasonable grounds for rejecting electrophoretic tests, even if the validity of this technique were accepted in general terms. As in the case of polygraphy, however, the more serious conceptual problem confronting the courts has been to decide what kinds of experts and what bodies of specialized knowledge should be consulted in operationalizing the standard of general acceptance.

It may seem surprising that this problem has risen in connection with electrophoretic blood typing, a technique that is universally accepted as being more securely grounded in scientific theory (genetics, molecular biology) than polygraphy. But three recent cases involving blood typing neatly illustrate why this fact by itself has not sufficed to establish the general scientific acceptability of electrophoresis.

In *State v. Washington*,[45] the first U.S. case to challenge the admissibility of electrophoretic bloodstain analysis, the court was confronted by conflicting testimony from experts of very different backgrounds and occupations. Dr. Grunbaum, a research chemist at the University of California and an expert on the design and use of electrophoretic equipment, testified for the defense that the particular form of electrophoresis used in that case, the multisystem method, was not reliable. By contrast, the chief witness for the prosecution, Mark Stolorow, a forensic scientist then employed by the State of Illinois Department of Law Enforcement,[46] testified that the multisystem method was reliable and was in widespread use at criminological laboratories around the country. The Supreme Court of Kansas weighed these conflicting accounts and concluded that the prosecution had provided sufficient evidence of reliability and general acceptance to justify admitting the technique.

The Michigan Supreme Court also reviewed the admissibility of electrophoretic testing on the basis of roughly similar evidence, but concluded by a slim majority, after two rounds of litigation, that the multisystem test was not reliable. In the first phase of *People v. Young*,[47] the Michigan court found the trial record inadequate and remanded the case to the lower court for a fuller hearing on the issue of admissibility. During this hearing, the defense, as in the Kansas case, presented Dr. Grunbaum as its chief witness. The prosecution called seven experts from both forensic and research laboratories, one of whom was Mark Stolorow. In the second phase of *Young*, the Supreme Court reviewed the record generated by the trial court and ruled that electrophoresis of bloodstains had not won general acceptance in the scientific community.[48]

The rejection of electrophoresis in Michigan reflected a fundamentally different judicial interpretation of the *Frye* standard from that embraced in *Washington*. While the Kansas court was content to regard both Grunbaum and Stolorow as qualified "experts," the Michigan court sought to differentiate between the defense and prosecution witnesses in more sociological terms. The Michigan judges began with the premise that general acceptance of electrophoresis can only be established by "disinterested and impartial experts in the particular field to which it belongs." Such experts, the *Young* majority further reasoned, could properly be drawn from a community of research scientists, but not of technicians. Applying this criterion, the court devalued the testimony of Stolorow and two other prosecution witnesses, all three of whom were relegated to the technician category. Their testimony was further discounted because two were full-time employees of law enforcement agencies and hence presumably not disinterested. Grunbaum, by contrast, was accepted as a scientist, probably because he held a Ph.D. in biochemistry, and the majority accordingly deferred to his judgment that the reliability of electrophoresis was not adequately established. A vigorous dissent characterized this analysis as "a regressive approach to scientific developments which, parenthetically, would have devalued the opinions of Jonas Salk, Albert Einstein, or Marie Curie."[49]

The third case in the series is *People v. Brown*,[50] in which the California Supreme Court refused to admit electrophoretic bloodstain typing but made no independent effort to assess the relative reliability of conflicting testimony. Instead, the court rested its decision on the fact that the trial record did not reveal the kind of scientific consensus required to establish reliability. The judges first reviewed the litigation in other jurisdictions in order to show that "the acceptance of tests for typing stale body-fluid stains is a matter of substantial legal controversy."[51] Under these circumstances, the court emphasized, the party offering the disputed evidence has the burden of proving its validity. In this case, the court held, there were several respects in which the prosecution had fallen short.

Borrowing a leaf from *Young,* the *Brown* court noted that two of the prosecution witnesses

> were competent and well-credentialed forensic technicians, but their identification with law enforcement, their career interest in acceptance of the tests, and their lack of formal training and background in the applicable scientific disciplines made them unqualified to state the view of the relevant community of *impartial scientists.*[52]

The court then conducted a brief review of the scientific literature, but only to show that "impartial science" had reached no clear consensus about reliability. To establish such a consensus, the court ruled, the trial court would have to develop a better record "with the help of live witnesses qualified in the applicable scientific disciplines."[53] As in *Frye,* however, the court gave no further direction for identifying the relevant disciplines.

We are now in a position to analyze the different types of boundary work employed in *Washington, Young,* and *Brown* for dealing with expert testimony. In the Kansas case, the court was reluctant to look behind the facade of claimed expertise to judge for itself whether any witness was inherently more qualified to speak on the disputed technical issues than any other. All were regarded indiscriminately as members of the "particular field" to which electrophoretic bloodstain analysis belongs. In the end, the court simply accepted the jury's judgment that the prosecution witness was more credible than the defense witness. This finding, in turn, led to the acceptance of bloodstain analysis as reliable.

In contrast, the majority in the second *Young* decision felt compelled to distinguish between scientists (presumed to be both competent and disinterested) and technicians (presumed to be neither competent nor disinterested) through a species of boundary work. By placing important prosecution witnesses on the technician side of the boundary, the court discounted their testimony and reached a negative conclusion about the reliability of the contested evidence.

The *Brown* court also engaged in boundary work, as noted above; however, its purpose in differentiating "scientist" from "technician" and "impartial" science from "interested" science was somewhat different from that of its Michigan counterpart. The California court used boundary work as an instrument for establishing lack of consensus but was not prepared to assert on this basis that electrophoresis was not generally accepted. Rather, the Supreme Court avoided the issue by delegating to a lower court the task of probing, with the help of witnesses from "applicable scientific disciplines," whether such a consensus exists.[54] What do these three very different strategies betoken for the future courtroom use of DNA fingerprinting?

DNA Fingerprinting

While the biological theories underlying DNA fingerprinting appear wholly uncontroversial, proponents of the technique are wary about other kinds of expert disagreements that may arise over attempts to move DNA-typing from the research laboratory to forensic medicine. As in the case of electrophoresis, one possible source of controversy is the unknown impact of environmental factors on the quality of the samples being analyzed, and hence on the accuracy of the tests. The electrophoresis cases serve warning that controversies over such subsidiary issues could lead to generalized judicial skepticism about the reliability of DNA fingerprints and to their eventual rejection as evidence.

Another factor that may prompt an adverse judicial reaction is the rapid commercialization of the DNA-typing technology. To date, only a small handful of private firms, such as Lifecodes and Cellmark, perform such services in the United States. These firms have also supplied the majority of expert witnesses to testify for the prosecution on the validity of the technique. The electrophoresis cases suggest that this class of experts may be especially vulnerable to discounting through judicial boundary work: not only might they easily be dismissed as technicians, but their affiliation with organizations that have a commercial stake in the new technology might make their testimony still more suspect. Accordingly, police departments may indeed do well to wait until consensus exists not only about the biological foundations of DNA fingerprinting, but about the reliability of collateral techniques for sampling, storage, and testing, before aggressively introducing such methods into criminal proceedings.

THE "SCIENCE" OF CLINICAL ECOLOGY

Polygraphy and DNA fingerprinting naturally lend themselves to judicial boundary work because both are applied techniques derived more or less persuasively from underlying scientific theories. Challenges to their reliability can thus be founded on the claimed deficiencies of either the science or the related technology. The resulting record invites judges to construct boundaries—necessarily subjective, intuitive, and arbitrary—between the testimony that they wish to regard as authoritative "science" and that which they choose to disregard as unscientific, unreliable, or merely technical.

But is this kind of response inevitable when novel scientific evidence enters the courtroom? One may wonder, for example, whether judges have equal scope for boundary drawing when the evidence before them consists of a new scientific theory rather than a novel instrument or technique. Judicial responses to testimony by practitioners of "clinical ecology" in recent toxic tort cases furnish a fruitful basis for further exploring this question.

Clinical ecology is a name used by a loosely organized community of immunologists and allergists to describe a set of beliefs concerning the way chemicals in the environment affect human health. Adherents of clinical ecology ascribe a wide variety of physical and psychic disorders (e.g., depression, chronic fatigue, respiratory and gastrointestinal symptoms, hypertension) to chemical exposure, alleging in particular that environmental chemicals produce "immune system dysregulation" in specially susceptible patients. Clinical ecologists have developed various treatment techniques that are claimed to be effective against chemically induced illnesses. These include major modifications in diet and in the patient's living and work environments, such as isolation in "safe" rooms with special air filtration systems.

Starting in the early 1980s, clinical ecologists began appearing as witnesses for plaintiffs in lawsuits against chemical producers and dischargers around the country. Their early appearances resulted in some notable victories for plaintiffs. For example, in *Anderson v. W. R. Grace*, testimony by Dr. Alan Levin, a leading practitioner of clinical ecology, led a federal district court in Massachusetts to deny the defendant's motion for summary judgment on the issue of causation. Levin testified that the cell damage he had observed in the plaintiffs was a precursor to leukemia and could have been caused by Grace's chemical discharges. Following its defeat on the summary judgment motion, W. R. Grace reached a multi-million dollar settlement with the plaintiffs.[55]

In a number of more recent cases, however, testimony by clinical ecologists has received much less hospitable treatment from the courts. Thus, in *Rea v. Aetna Life Insurance Company*, a U.S. district court in Texas dismissed an antitrust suit by a group of clinical ecologists, finding *inter alia* that "clinical ecology is not a board certified specialty of medicine requiring formal training or testing nor identifiable with any set of independently established standards of practice or body of knowledge."[56] Similarly, in *Laborde v. Velsicol Chemical Corp.*, a Louisiana court held that " 'Clinical Ecology' is not a recognized field of medicine."[57] Finally, in *Sterling v. Velsicol*,[58] the Sixth Circuit Court of Appeals held that the opinions of the plaintiffs' experts on immune system dysregulation, including testimony by Dr. Levin, lacked sufficient scientific foundation to prove that water contaminated by the defendants had injured the plaintiffs.

What accounts for this radical reversal in judicial receptivity toward clinical ecology? The answer lies partly in the way the medical community reacted to this self-proclaimed specialty between 1980 and the present. During this period, several professional organizations concerned with allergies and immunology reviewed and criticized the scientific basis for the claims of the clinical ecologists. In 1985, the influential American Academy of Allergy and Immunology (AAAI) drafted a position statement concluding that

The theoretical basis for ecologic illness in the present context has not been established as factual, nor is there satisfactory evidence to support the actual existence of "immune system dysregulation" or maladaptation.[59]

AAAI therefore labeled clinical ecology "an unproven and experimental methodology."

Another powerful professional organization, the California Medical Association (CMA), had adopted the position as early as 1981 that clinical ecology did not constitute a valid medical discipline. The increasing publicity accorded to this specialty led CMA in 1984 to convene a scientific task force to reevaluate the question in detail. Proceeding with its appointed business involved the task force in several types of professional boundary work. To begin with, members had to be selected from disciplines competent to comment on clinical ecology. The CMA task force elected to be broadly representative of the medical profession, choosing members "for their expertise in internal medicine, toxicology, epidemiology, occupational medicine, allergy, immunology, pathology, neurology and psychiatry."[60] So diversely constituted a body could claim with some credibility to be speaking "for medicine." Additional boundaries were drawn as the task force determined what criteria it should apply in judging the validity of the clinical ecology literature. These methodological criteria—the product of interdisciplinary consensus within the task force—implicitly established the boundary between legitimate and illegitimate medical science.

Interestingly, the CMA task force conducted not only a scientific proceeding (a literature review) but a quasi-judicial one as well. A public hearing was held in April 1985, at which both proponents and opponents of clinical ecology presented oral testimony to task force members. Based on these investigations, the task force concluded that there was no convincing evidence "that patients treated by clinical ecologists have unique, recognizable syndromes, that the diagnostic tests employed are efficacious and reliable or that the treatments used are effective."[61]

These efforts at consensus building within the medical community had a significant impact on the courts. The *Velsicol* opinion provides the clearest illustration. Referring to the activities of AAAI and CMA, the court noted that "the leading professional societies in the specialty of allergy and immunology . . . have rejected clinical ecology as an unproven methodology lacking any scientific base in either fact or theory." Other medical organizations, the court conceded, had not yet completely discredited

clinical ecology, but neither had they endorsed its scientific methodology or experimental findings. Moreover, the plaintiffs' experts
had supplied no studies demonstrating the effects of Velsicol's
chemical discharges, carbon tetrachloride and chloroform, on the
immune system.

It is worth noting that the *Velsicol* court made no attempt to look
behind the consensus generated by AAAI and CMA to determine
whether these organizations possessed any hidden biases. As far as
the court was concerned, both professional societies could simply
be accepted as "black boxes" capable of distinguishing sound medical theories from unsound ones. That established medical societies
might not be wholly disinterested in their efforts to discredit a
competing, and apparently successful, specialty does not seem to
have worried the *Velsicol* court.[62]

The opinion contains just one suggestion that a different strategy
on the plaintiffs' part might have persuaded the court to look more
skeptically at the statements issued by AAAI and CMA. The
clinical ecologists who testified for the plaintiffs were not their
treating physicians. In dismissing the plaintiffs' contentions about
immune system dysregulation, the court noted that these experts
had "neither personally examined or interviewed plaintiffs, nor
performed the requisite medical tests." Without such clinical tests
and a widely accepted medical basis for their conclusions, the court
held that "plaintiffs' experts' opinions are insufficient to sustain
plaintiffs' burden of proof that the contaminated water damaged
their immune system."

Here, then, is evidence of an independent judicial bias that could,
under appropriate circumstances, lead courts to draw boundaries
conflicting with an apparent scientific or medical consensus. In
cases involving personal injury claims, courts have traditionally
shown greater deference to the plaintiff's treating physician than to
experts who were not personally familiar with the plaintiff's medical history. This bias betrays a preference for empirical evidence—
one might even say "eyewitnessing"—that many would view as
incompatible with the cast of modern science. Yet courts have
repeatedly confirmed that they are prepared to trust otherwise
questionable medical testimony so long as it is presented by a
treating physician.

In *Menendez v. Continental Insurance Company*,[63] for instance,
a Louisiana court upheld the general rule that, as a matter of law,
courts should give greater weight to the treating physician than to
other experts. The case provides an interesting and contrasting

footnote to *Velsicol*, because the plaintiff's principal witness, Dr. Alfred Johnson, was both her treating physician and a practitioner of clinical ecology. Having accepted Dr. Johnson as a treating physician, the *Menendez* court deferred to his judgment that the plaintiff's condition had been "triggered" by an adverse drug reaction

> that "sensitized" plaintiff to many things in her environment, including formaldehyde found in carpet, cigarette smoke, and such foods as carrots, sweet potatoes, chicken, walnuts, rice, beef, eggs, certain fruits, kidney beans, soy, red snapper, salmon, shrimp and flounder.[64]

The defendant's expert countered this testimony by asserting that "he did not think clinical ecology was a scientific practice," but the appeal to science did nothing to change the Louisiana court's legally driven construction of the relative credibility of the two experts.

CONCLUSION

The examples discussed in this essay reveal some general patterns in the way that judges think about science and technology, especially when confronted by new developments in these fields. Both the polygraphy and the electrophoresis cases indicate considerable judicial sympathy for an idealized conception of "science." Judges consistently defer to testimony that conforms to their notion of the norms of science, particularly the norm of disinterestedness or impartiality.[65] As judges assess the credibility of conflicting experts, they engage in boundary work that serves not only as an expedient for resolving expert conflict but also as a strategy for keeping intact the ideal image of science. Judicial boundary drawing between "scientists" and "technicians" is an especially telling example of this technique.

The application of legal decision making techniques to new scientific evidence discloses an interesting paradox. On the one hand, judges seem inclined to uphold the cognitive authority of science through boundary work resembling that done by scientists themselves. This approach permits courts to legitimate decisions about the admissibility or reliability of technical evidence with authority borrowed from science. On the other hand, courts are also prepared, when the occasion demands, to evaluate experts and evidence according to legal norms that have no validity in science. In the process, they may draw boundaries that are unacceptable or incomprehensible to scientists, but that reaffirm the separate, even superior, normative authority of the law. The treating physician rule provides a striking example of this approach.

Of course, even when courts seem to defer to the authority of science, the construction of knowledge in the courtroom usually proceeds in accordance with legal, rather than scientific, rules of the game. The adversarial and case-specific nature of most lawsuits produces ony a partial picture of the science relevant to the legal dispute. In deciding questions of admissibility, courts are usually confined to the testimony of a few selected witnesses claiming to represent much larger professional communties. Moreover, courts have never found it easy to guard against "professional witnesses" (Grunbaum, Stolorow, and Levin, for example), who have been specially selected for their ability to represent extreme, idio-syncratic, or biased scientific viewpoints. Short of converting them-selves into mini-administrative agencies—as the Michigan and California trial courts were asked to do in the electrophoresis cases—courts have little opportunity to review any field of science as a whole.

These structural limitations go a long way toward explaining why courts so frequently disagree with each other and with the scientific community at large over issues of admissibility. Ultimately, the construction of scientific knowledge in the courtroom responds to the highly particularized needs of the disputing parties and to the legal system's general interest in closure. A process driven by this nexus of interests can hardly be expected to produce the same results (i.e., certify the same "facts") as the processes of knowledge construction in conventional scientific forums.

The fate of clinical ecology in recent judicial decisions suggests, however, that concerted action by scientific and technical commu-nities sometimes helps their collective views to win respect from the courts. The strategy followed by AAAI and CMA in discrediting clinical ecology proved effective because those organizations de-fined the boundaries of valid medical science in a way that courts could not readily afford to ignore. Marshaling their professional authority with considerable skill, both organizations declared that clinical ecology was not a generally accepted medical specialty and, by implication, that practitioners of clinical ecology were not legiti-mate members of the medical profession. As we have seen, the development of this consensus achieved the desired legal effect.

Judges began questioning the credibility of clinical ecologists testifying in toxic tort cases, and these witnesses accordingly ceased to play an important part in constructing science for the courts. With the clinical ecologists out of the picture, the conclusions of the two medical societies were imported more or less unchallenged into the law.

To summarize, the case law on new scientific evidence surveyed in this essay gives scientists and technical experts limited grounds for optimism. The example of clinical ecology suggests that timely initiatives by a community of experts can preempt the social construction of science that ordinarily takes place in the courtroom. In such cases, courts have little choice but to go along with the prevailing expert consensus about the acceptability of the evidence. But the decentralized institutions of science and technology will rarely take such action in the absence of extraordinary threats to their authority and autonomy. As the continuing muddle over polygraphy shows, controversies over admissibility are not about to become an endangered legal species.

NOTES

1. See, for example, Bruno Latour and Steve Woolgar, *Laboratory Life: The Construction of Scientific Facts* (Princeton: Princeton University Press, 1986).

2. Branden B. Johnson and Vincent T. Covello, eds., *The Social and Cultural Construction of Risk* (Dordrecht: Reidel, 1987).

3. Wiebe E. Bijker, Thomas P. Hughes, and Trevor J. Pinch, *The Social Construction of Technological Systems* (Cambridge: MIT Press, 1987).

4. In the United States, law-science interactions have been analyzed primarily within the highly specialized substantive and stylistic confines of the traditional law review article. A recent and welcome addition to the law-science literature from a more sociological perspective is Roger Smith and Brian Wynne, eds., *Expert Evidence: Interpreting Science in the Law* (London: Routledge, 1989).

5. See, for example, Edmund W. Kitch, "The Vaccine Dilemma," *Issues in Science and Technology* 2 (1986): 108–21.

6. See, for example, Peter Huber, "Safety and the Second Best: The Hazards of Public Risk Management in the Courts," *Columbia Law Review* 85 (1985): 277–337.

7. Thomas F. Gieryn, "Boundary-Work and the Demarcation of Science from Non-Science," *American Sociological Review* 48 (1983): 781–95. Other social actors besides scientists have also found it necessary to engage in boundary work distinguishing between science and non-science. These include policymakers, industry, and the news media. See, for instance, Sheila Jasanoff, "Contested Boundaries in Policy-Relevant Science," *Social Studies of Science* 17 (1987): 195–230.

8. 293 F. 1013 (D.C. Cir. 1923).

9. 293 F. at 1014.

10. Office of Technology Assessment, *Scientific Validity of Polygraph Testing* (Washington, D.C.: Office of Technology Assessment, 1983), p. 30; David T. Lykken, *A Tremor in the Blood* (New York: McGraw-Hill, 1981), p. 218.

11. Paul C. Gianelli, "The Admissibility of Novel Scientific Evidence: *Frye v. United States,* a Half-Century Later," *Columbia Law Review* 80 (1980): 1197.

12. Faust E. Rossi, "Modern Evidence and the Expert Witness," *Litigation* 12 (1985): 20.

13. Ibid.

14. M. J. Saks and R. Van Duizend, *The Use of Scientific and Technological Evidence in Litigation: Report and Indexed Bibliography* (Williamsburg, Va.: National Center for State Courts, 1983).

15. State v. Williams, 446 N.E. 2d 444, 448 (Ohio 1983).

16. M. Berger, "A Relevancy Approach to Scientific Evidence," *Jurimetrics* 26 (1986): 245.

17. Federal Rule of Evidence 403.

18. 463 U.S. 880 (1983).

19. Brief Amicus Curiae for the American Psychiatric Association in Barefoot v. Estelle, March 4, 1983.

20. 463 U.S. at 896.

21. 463 U.S. at 902 (approvingly quoting language from the trial court opinion).

22. United States v. DeBetham, 348 F.Supp. 1377, 1383 (S.D. Cal. 1972). See also Mark Elliott Squires, "The Truth About the Lie Detector in Federal Court," *Temple Law Quarterly* 51 (1978): 78–79.

23. Marston's exploits include some that were distinctly out of character for a Ph.D. from Harvard. Immediately after the Lindbergh kidnapping, for example, he tried in vain to get his invention accepted by Col. Lindbergh, Bruno Hauptmann's defense counsel, and the governor of New York. A later invention proved more marketable: under the pseudonym of "Charles Moulton," Marston created the comic strip character Wonder Woman, who is famed for using a still more primitive technology than the Marston lie detector (a rope) to make her opponents tell the truth.

24. Lykken, *A Tremor in the Blood,* pp. 27–30. Marston's original invention detected changes in the subject's systolic blood pressure, whereas Larson's machine recorded continuous changes in heartbeat, respiration, and blood pressure.

25. United States v. Alexander, 526 F.2d 161, 163 (8th Cir. 1975).

26. 293 F. at 1014.

27. 348 F.Supp. 1377 (S.D.Cal. 1972).

28. United States v. Wilson, 361 F.Supp. 510 (D.Md. 1973).

29. United States v. Urquidez, 356 F.Supp. 1363 (C.D.Cal. 1973).

30. Interrogation methods, for instance, have undergone major improvements since Larson developed the relevant/irrelevant technique, which involves asking the subject a mixture of questions relevant and irrelevant to the inquiry. For an account of the methods in common use today, see Lykken, *A Tremor in the Blood.*

31. Squires, "The Truth about the Lie Detector," p. 82.

32. Lykken, *A Tremor in the Blood,* pp. 64–68.

33. United States v. Ridling, 350 F. Supp. 90 (E.D.Mich. 1972).

34. 526 F. 2d 161, 164 (8th Cir. 1975).

35. Alfred Meyer, "Do Lie Detectors Lie?" in *Criminal Justice 84/85* ed. Donald MacNamara (Guilford, Conn.: Dushkin Publishing Group, 1983), p. 157.

36. Squires, "The Truth about the Lie Detector," p. 89.

37. Commonwealth v. Vitello (1978).

38. "Court Decides Prosecutor Can't Veto Lie Test," *Boston Globe*, 23 April 1987.

39. Jean L. Marx, "DNA Fingerprinting Takes the Witness Stand," *Science* 240 (1988): 1616.

40. This now standard technique of molecular biology involves the use of restriction enzymes to cut strands of DNA into shorter fragments at particular sites.

41. Electrophoresis is the technique of using an electrically charged field to separate differently charged molecules. In the case of DNA fingerprinting, separation occurs because the smaller fragments of DNA move through the field faster than the larger ones.

42. See, for example, Kirk Johnson, "DNA 'Fingerprinting' Tests Becoming a Factor in Courts," *New York Times*, 7 February 1988.

43. David D. Dixon, "The Admissibility of Electrophoretic Methods of Genetic Marker Bloodstain Typing under the *Frye* Standard," *Oklahoma City University Law Review* 11 (1986): 777.

44. See, for example, People v. Brown, 40 Cal.3d 512, 534 (1985).

45. 229 Kan. 47 (1981).

46. 229 Kan. at 51.

47. 418 Mich. 1 (1983).

48. People v. Young, 391 N.W.2d 270 (1984).

49. 391 N.W.2d at 290.

50. 40 Cal.3d 512.

51. 40 Cal.3d at 532.

52. 40 Cal.3d at 533.

53. 40 Cal.3d at 535.

54. California trial courts have subsequently used the discretion granted them by the Supreme Court in *Brown* to hold that electrophoretic blood typing is reliable. See, for example, People v. Reilly, 242 Cal. Rptr. 496 (Cal.App. 1 Dist. 1987).

55. Laurie A. Rich, " 'No Winners' in an $8–9 Million Settlement," *Chemical Week*, 8 October 1986, pp. 18–19.

56. Civ. Action No. 3-84-0219-H (N.D. Tex. 1985).

57. Sheridan and Tupi, "Joint Arrangement Results in Victory for Chemical and Insurance Companies," 332 PLI/Lit 447 (1987).

58. 855 F.2d 1188 (6 Cir. 1988).

59. American Academy of Allergy and Immunology, *News and Notes*, Summer 1985, p. 11.

60. California Medical Association Scientific Board Task Force on Clinical Ecology, "Clinical Ecology—A Critical Appraisal," *Western Journal of Medicine* 144 (1986): p. 240.

61. Ibid., p. 243.

62. Within science, activities of the sort undertaken by AAAI and CMA to read clinical ecologists out of the profession are by no means uncommon. Critics of science have ordinarily interpreted such actions as maneuvers to maintain the authority of scientific disciplines or organizations. In short, such boundary drawing usually reflects professional self-interest. See, for example, Thomas F. Gieryn and A. E. Figert, "Scientists Protect Their Cognitive Authority: The Status Degradation Ceremony of Sir Cyril Burt," in *Sociology of the Sciences Yearbook* 10 (1986), special number edited by G. Bohme and N. Stehr.

63. 515 So.2d 525 (La.App. 1 Cir. 1987).

64. 515 So.2d at 527.

65. This is, of course, one of the four elements of the scientific ethos singled out by Robert Merton in his classic essay on the norms of science. See Robert K. Merton, "The Normative Structure of Science," reprinted as Chapter 13 in Robert K. Merton, *The Sociology of Science* (Chicago, Ill.: University of Chicago Press, 1973), pp. 267–68. It would be interesting, though beyond the scope of this essay, to investigate whether the judicial concept of science conforms perfectly to the Mertonian norms.

The Nuts and Bolts of Democracy: Toward a Democratic Politics of Technological Design

RICHARD E. SCLOVE

Technology, it seems, is a taboo subject in the social sciences. This is especially true of the research and development (R&D) process. Yes, there is an occasional passing mention; even a seminal book or two—especially in the history of technology, as I acknowledge in my endnotes. But the mainstream of social science and philosophy proceeds as though concrete technologies and their professional practitioners simply do not exist, let alone matter.

The principal exception occurs within the discipline of economics, in which a modest scholarly industry now preaches the need to devise public policies that can accelerate the pace of technological innovation.[1] The stated objective of such policies is to enhance national economic growth, productivity, and international competitiveness—on the unexamined assumption that as long as an innovation sells profitably, it is an unalloyed social blessing.

The situation is astonishing. It is as perplexing as if for generations a family shared its home with a giant pink epileptic elephant, and yet never discussed—somehow never even noticed—the beast's presence and pervasive influence on their lives. It is even as though everyone in the United States gathered together each night in their dreams—assembled solemnly in a glistening moonlit glade—and there furiously debated, radically rewrote, and ratified a new constitution. Then, awakening with no memory of what had passed, they yet spent each next day mysteriously complying with the nocturnally revolutionized document in its every word and letter.

Such a wildly impossible world, in which unconscious collective actions govern waking reality, is the world that now exists.[2] It is the modern technological world that we have all—albeit some more potently or self-seekingly than others—helped create. Hence my central message: we must learn to take the title of this essay not only seriously, but literally. The *nuts and bolts of democracy*—ordinarily a metaphor through which we denote concern with the practical arrangement and governance of a democratic society—must grow to encompass a literal concern with nuts and bolts. An engaged citizenry must become critically involved with the choice, governance, and even design of technological artifacts and practices, and committed specifically to adopting only those technologies which are themselves compatible in their design with reproducing over time the society's democratic nature. Or else, like hapless daytime victims of our own midnight cabals, we can have no democracy.

I have elsewhere provided philosophical justification for this claim.[3] Here I want to begin to explore its practical entailments.

TECHNOLOGY AND DEMOCRACY

A key premise of my argument is the growing recognition that technologies represent an important species of social structure, functionally comparable to other sorts of social structures such as dominant political and economic institutions, laws, or systems of cultural belief.[4] Each of these—like technologies—qualify as social structures by virtue of being contingent social products that, once they come into existence, tend to endure and profoundly influence (without, however, fully determining) one another's subsequent evolution, as well as the course of history and the texture of daily experience.[5]

But what has this insight got to do with democracy, and what is so important about democracy anyway? Simply put—and here I rely upon a rational reconstruction of the political philosophies of Rousseau and Kant—I believe that democracy is a necessary background condition for enabling people to develop individual autonomy and to decide together what should matter to them.[6] Democracy is in this sense a highest-order shared human value, because it is fundamental to formulating and realizing other values.

I suggest that we as individuals have a moral obligation to help establish democratic social orders whose institutional essence should be to provide members with equal and extensive opportunities to determine the collective conditions of their existence. However, an important a priori constraint upon such political involvement is that citizens must reproduce over time the democratic nature of their society.

Implications for technological design and practice emerge straightforwardly. If the essence of democracy is that citizens ought to be equally empowered to participate in determining the collective conditions of their existence, particularly as these bear upon the possibility of democracy itself, and if technologies are themselves freedom-conditioning social structures, it follows that it is normally imperative to democratize technological design and practice.

What does this mean? Technological democratization has both procedural and substantive components. Procedurally, there must be expanded opportunities for people from all walks of life to participate effectively in guiding the evolving technological order. Substantively, the resulting technologies ought to be compatible in their design with democracy's necessary conditions. (For specific design recommendations, see table 1.)

Mine is a simple argument. With respect to contemporary technological trends, its implications are also thoroughly radical. Assuming its validity, then it is morally imperative—a necessary condition of our own freedom and dignity—to take processes of technological development that are today guided primarily by market forces, economic self-interest, or international rivalry, and instead subordinate them to democratic prerogatives.

Notice that this does not mean that technologies are necessarily the single most important influence on political life. Rather, they are sufficiently important—and so inextricably intertwined with other such factors—that we must learn to subject them to the same rigorous political scrutiny and involvement that we accord to other influences—such as laws, international relations, and the distribution of wealth and power.

Just as technology is not the only important consideration that bears upon democracy, it is also not true that democracy is the only important consideration in our technological decision making. Nevertheless, we should treat democracy as a first-order consideration because—as I have said—it provides the necessary background for being able to decide freely and fairly what other considerations to take into account.

TOWARD A DEMOCRATIC POLITICS OF TECHNOLOGY

This is not the place to lay out a comprehensive program for institutionalizing a democratic politics of technology. Space is lim-

Table 1
A Provisional System of Rationally Contestable Design Criteria for Democratic Technologies and Architecture

Toward Democratic Community:

A. Seek Balance between Communitarian/Cooperative, Individualized, and Inter-Community Technologies. Avoid Technologies that Establish Authoritarian Social Relationships.
B. Seek Technologies that Can Help Enable Disadvantaged Individuals and Groups to Participate Fully and Autonomously in Social and Political Life.

Toward Democratic Work:

C. Seek Equal and Extensive Availability of a Diverse Array of Flexibly Schedulable, Self-Actualizing Technological Practices. Avoid Meaningless, Debilitating, or Otherwise Autonomy-Impairing Technological Practices.

Toward Democratic Politics:

D. Insofar as Technologies Each Tend to Embody or Convey Symbolic Meaning, Avoid Technologies that Promote Impoverished or Otherwise Distorted Human Understanding and Self-Understanding.

To Help Secure Meaningful Collective Self-Governance:

E. Seek Relative Local Economic Self-Sufficiency.
F. Seek to Restrict the Distribution of Potentially Adverse Consequences (e.g., environmental or social harms) to Within the Boundaries of Local Political Jurisdictions.
G. Seek Technologies (including an Architecture of Public Space) Compatible with Egalitarian, Globally Aware Political Decentralization and Federation.

To Help Reproduce Social Structures Over Time:

H. Seek Ecological Sustainability.
I. Seek Technologies that Are Relatively Invulnerable to Catastrophic Sabotage and to the Attendant Risks of Civil Liberties Abridgment.
J. Seek "Local" Technological Flexibility and "Global" Technological Pluralism.

For the derivation of these design criteria see Richard E. Sclove, "Technology and Freedom: A Prescriptive Theory of Technological Design and Practice in Democratic Societies" (Ph.D. diss., Massachusetts Institute of Technology, 1986), pp. 491–777.

ited, variability in contextual circumstances dictates variability in tactics and objectives, and in the short run the greatest need and opportunity may be for countless creative initiatives at the local level. But to help stimulate discussion and requisite audacity, provide a little more concreteness, and launch some initial action, I offer a few suggestions.[7]

I have tried to list these recommendations in an order that roughly reflects the descending feasibility of implementing them on a short-run and strictly local basis. That is, groups of citizens, civic and other organizations (including civic-minded churches, unions, or businesses), universities, or state and local governments could begin to undertake those nearer the top of the list now, one recommendation at a time, and independently of actions taken or not taken by other communities, organizations, or levels of government. Conversely, to carry out the recommendations nearer the bottom of the list would presumably require a progressively greater degree of translocal political interest and coordination.

It is important to bear in mind, however, that working through formal governmental procedures or established political parties is only one among a variety of possible, complementary routes to technological democratization. Thus even small groups of citizens, capitalizing on any skills and experience in democratic leadership that happen to be at hand, can begin immediately to grasp, create, and expand opportunities for self-empowerment in technological politics.[8]

Beginning now to pursue those steps that can feasibly be implemented at a local or state level is entirely worthwhile in itself. In terms of political strategy, it would also help generate the broader political will or movement that institutionalizing further-reaching recommendations would require. Gradual implementation of the latter would, in turn, facilitate, extend, and consolidate any ongoing local efforts, which could themselves conceivably evolve into regional, nationwide, or transnational efforts. But regardless of the precise arenas of action, the overarching goal remains the same: to help recreate a societywide culture of participation and collective self-determination, accompanied by a deepening commitment to evolving technologies and other social structures that are substantively democratic.

These are my specific recommendations:

Reconnaissance of Needs

Organize community oral history projects to help disclose the historical pattern and causes of local technological change. Then initiate community self-assessments of the technological order's present degree of compatibility with the necessary conditions of democracy, and of needs and opportunities for improvement. This could provide a context for beginning to debate publicly the formulation and use of design criteria for democratic technologies (see table 1). Good in itself, participation in the debate of design criteria would also help citizens learn to better perceive and anticipate technologies' latent political and cultural dimensions.

Education and Research for a Democratic Technological Order

Education: Create new school, college, and continuing-education curricula oriented toward increased popular understanding of the social and political implications of technology, and toward increased citizen competence to participate in all phases of democratic technological politics, including design. The courses and publications on social control of technology that have been developed under the auspices of Britain's Open University provide an instructive example.[9]

Research: Seek financial support for new programs of research and graduate assistantships oriented toward the development of democratic technologies and architecture. Seek support also for strengthened social scientific research—conducted collaboratively by lay citizens and scholars—into the social history and anthropology of technology, patterns of sociotechnological dynamics, and the design and evaluation of projects for advancing technological democratization. The Highlander Research and Education Center in Tennessee is one group that has pioneered in facilitating participatory research projects.[10] Collaborate with research programs that are underway in other countries, including socially or technologically innovative programs in Scandinavia, eastern Europe, and the Third World.

Publicity: Supplement education and research with programs of literary and artistic popularization of efforts around the world to create democratic technologies and to expand citizen opportunities to participate in technological decision making. Encourage print and broadcast media to run documentaries, regular columns, guest editorials, and op-ed pieces focusing on social criticism of technology and technological politics. Also explore other novel means for interactively sharing knowledge and ideas.

Political Movements

Attempt to integrate the concept of democratic technologies into the agenda and practice of progressive political movements and organizations such as environmental groups, progressive labor unions, the National Organization for Women, the NAACP, the American Civil Liberties Union, and High Technology Professionals for Peace.

Participation in R&D and Design

Strive to generate substantive opportunities for citizen involvement in technological research, development, design, and strategic planning within universities, corporations, and government laboratories and research and planning agencies. The Lucas Aerospace workers and London's technology networks (to be discussed later), along with architectural community design centers, provide several illustrative models.[11]

Time for Politics

Begin to implement or extend programs of job sharing, flexible life-scheduling, child care, and citizen sabbaticals. In addition to supporting individual self-actualization (see table 1, criterion C), this would expand citizens' opportunities to participate in politics, including technologial politics and design.

Relative Local Economic Self-Sufficiency

Promote institutional conditions and technologies favorable to local production for local consumption (again see table 1, especially criterion E). If prudently devised, such an economic strategy can expand both the number and diversity of jobs, contribute to ecological sustainability (e.g., less nonrenewable energy would be expended in transportation), enhance the possibility of meaningful local self-governance, and provide a buffer against the untoward shocks of excessive integration into transregional and global markets. In the United States, efforts in this direction have been pioneered by organizations such as the Institute for Local Self-Reliance in Washington, D.C.; the Rocky Mountain Institute in Snowmass, Colorado; and the New Alchemy Institute in East Falmouth, Massachusetts.[12]

Democracy and Social Impact Statements (DSIS)

Seek legislation, at local, state, or national levels, that would require corporations and government agencies to file a public DSIS before introducing a technological innovation. A DSIS should consider both short-run impacts and impacts that might occur if the technology and its supporting infrastructure become widely deployed.[13] Each new kind of technology or design innovation should be accompanied by a generic DSIS that delineates both potential and probable impacts within all reasonably anticipatable contexts of production, application, and use. In addition, construction of large technological facilities, networks, or systems—for example, a factory, shopping mall, highway, central station electrical plant, or telecommunications system—should be preceded by an individual or nongeneric DSIS. A representative group of potentially affected citizens, who are neither technically expert nor members of the organization introducing the innovation, should oversee the production of each DSIS. The specific function of DSIS's could range from merely informational to providing a basis for various forms of political oversight or regulation.

Let me add one caution. We should learn to evaluate with extreme wariness optimistic claims that a particular technology will do much to enhance democracy. Historically, such claims have normally been based upon treatment of a mere possibility as though it were a contextual probability, while overlooking nonfocal consequences whose aggregate social significance often swamps that of technology's focal purpose.

Office of Technology Assessment (OTA)

Require that the congressional OTA include technologies' degree of compatibility with democracy as a highest-order evaluative consideration. Beyond studying one kind of technology at a time, OTA should undertake studies of the interactive social implications generated by *complexes* of technologies. Have the OTA continue its earlier experiments with lay participation in technology assessment,[14] including the running of trials with lay technology assessment panels. The OTA should compile reports of the community self-assessments recommended earlier and report to the Congress on special needs, threats, and opportunities bearing on the possibility of creating a more democratic technological order. Create a national hot line ("1-800-BAD-TECH"?) through which citizens, including government and corporate employees, can file warnings of potentially antidemocratic technological developments.

Democratize Corporations

Provide incentives—or simply mandate—that corporations implement programs of transition to greater worker self-management, including worker participation in research and strategic planning. Require substantial worker and community representation on corporate boards. Based in part on recommendations from regulatory agencies and the OTA, the Congress or a designated agency should assess a special tax against corporations that fail to democratize at a reasonable pace or that persist in producing democratically harmful technologies. Tax incentives or subsidized loans should be available to assist the start-up of businesses that are democratically favorable.[15]

Civil Technological Empowerment

Communities must establish a capacity to respond to their identified needs for a more democratic technological order. When needed, there should be state or national assistance in helping localities plan, finance, and implement: the recreation of public space; the discouragement, dismantling, or replacement of democratically adverse technologies; the solicitation, sponsorship, or undertaking of necessary social science or hardware-oriented R&D; and the deployment of civic technologies and architecture supportive of democratic community and local and translocal democratic politics.

National Laboratories, the Pentagon, and Defense Industries

Recycle the national laboratories' research agenda away from their nearly exclusive emphasis on military applications. Initiate new research programs oriented toward developing civil technologies that would help support a democratically sustainable global future. Establish one or more national laboratories dedicated specifically to democratic technology. In addition to undertaking social scientific and hardware-oriented research, these labs should serve as information clearinghouses. To whatever extent the present national concept and practice of "defense" may be both domestically and internationally hurtful to democracy,[16] we should challenge and reform both.

Democracy, Multinational Corporations, and the Global Economy

Certain forms of corporate enterprise—notably large multinationals, international banks, and multipurpose conglomerates— may be too large to be effectively democratized internally, and too powerful to be effectively subordinated to external democratic guidance. We should identify such corporate forms and seek their devolution toward democratically manageable alternatives. A first step might be to establish a presidential commission with a mandate to issue recommendations on ways of transforming the present multinational corporate economy into one that can indeed by subjected to democratic prerogatives. We should also reorient the primary justification for government antitrust activity away from a purely economic concern to prevent monopolistic pricing, in favor of the broader-reaching concern that large corporations acquire democratically unjustified political and economic power.[17]

PRACTICAL EXAMPLES

Any of the preceding recommendations could, of course, be refined and made more compelling through elaboration. I will take, as one example, participation in R&D and design. One obvious objection might be that citizens, including workers, who are not technically expert could never participate meaningfully in the esoteric worlds of technological development or architectural design.

A brief review of case studies in diverse social domains can shed further light on this issue, as well as indicate additional opportunities and obstacles for lay participation in R&D and design.

Worker and Citizen Design of Socially Responsible Technologies

Consider a sequence of events that took place in Britain. In the late 1960s the Lucas Aerospace Company sought to lower fixed costs by closing some factories and laying off workers. In response the workers, who were organized in a company wide combine of unions that included both shop floor personnel and highly trained engineers, proposed to maintain employment by converting the factories to the manufacture of a new range of products. Soliciting ideas and assistance from universities and local communities but relying substantially on the latent creativity of all ranks of Lucas workers, the union developed a six-volume alternative corporate plan that satisfied several new criteria for women. These encompassed a commitment to producing more socially useful goods—rather than, say, military jet engines—and to do so by more democratic work processes, including more creative tasks for workers. The Lucas Aerospace Workers' Plan eventually included a hundred and fifty product ideas, accompanied by economic and engineering analysis, and some working prototypes. The ideas ranged from inexpensive medical devices to solar energy hardware, fuel-efficient vehicles, multifuel/multiuse electric generators adapted to Third World village needs, and special robotic devices that would enable workers to exercise their skills at a safe distance from hazardous work environments.

The proximate tactical results were not encouraging. The company toyed with several of the new products but rejected outright the basic concept of worker and community participation in design, innovation, and strategic planning. Lucas also managed—after Margaret Thatcher's Conservative Party took power—to fire union leaders. Yet even these events were not without value: they disclosed some of the political-economic barriers to technological democratization, and provided a springboard for follow-on endeavors.

Through public talks, articles, books, radio interviews, television documentaries, and incorporation into the televised curriculum of a popular open university continuing-education course on social control of technology, the Lucas saga has spread widely. Similar alternative corporate plans have been developed at other British companies and in other nations, including Australia. Collaborating with North East London Polytechnic, Lucas workers have designed a fuel-efficient bus that can travel on both roads and rail. The aim is to demonstrate a promising transportation technology while carrying to local communities an on-board multimedia presentation of the Lucas story. The purpose is to stimulate further political and technological collaboration between local communities and socially minded workers and engineers.[18]

Meanwhile, a former leader in the Lucas union, Mike Cooley, has become director of the Greater London Enterprise Board's (GLEB's) Technology Division, which has set up a series of technology networks. These are technically equipped workshops with access to nearby college libraries and faculty, where citizens can come for help in developing and marketing socially responsible technologies. GLEB has also begun a program of offering start-up subsidies to small businesses that meet specific social criteria, such as worker self-management or minority employment. More than two hundred enterprises have been initiated, with the technology networks functioning as a participatory resource for their research and development needs.[19]

Democratic Design of Workplace Technology

In Scandinavia, unionized newspaper graphics workers—in collaboration with sympathetic university technical researchers, a

Swedish government laboratory, and a state-owned publishing company—have succeeded in inventing a new kind of computer software. Instead of following recent trends toward routinized or mechanized newspaper layout, it enable printers and graphic artists to continue to exercise considerable creativity in page design. Known as UTOPIA, this project is an instance of broadened participation in the R&D process leading to a design innovation that, in turn, supports one condition of democracy: creative work. (See table 1, criterion C.)

UTOPIA is less ambitious in overall scope than the Lucas Aerospace workers' attempt, which sought not only to democratize the work process but, in addition, to produce more socially useful products. But UTOPIA has gone further practically, by moving beyond the development of prototypes to a successfully used design innovation. This instance of collaboration between workers and technical experts—so far limited to a single technology within a single industry—occurred under unusually favorable social and political conditions. Sweden's workforce is 85 percent unionized. And the nation's prolabor Social Democratic party has held power during most of the past half century, providing a supportive legislative context.[20]

Participatory Architecture

Compared with the relative paucity of attempted exercises in participatory design of machinery, appliances, or technical infrastructure, there is a rich history of citizen participation in architectural design. The range of stories is extremely diverse, including cases in which it proved difficult to motivate participation or in which the design outcome was little different from what an architect working alone might have devised. But in other cases highly novel designs resulted—sometimes evoking resistance from powerful social interests and institutions.[21]

One example is the Zone Sociale at the Catholic University of Louvain Medical School in Brussels. Here in 1969 students insisted that new university housing mitigate the alienating experience of massive hospital architecture. Architect Lucien Kroll established an open-ended, participatory design process that elicited complex organic forms, richly diverse patterns of social interaction—for example, a nursery school situated near administrative offices and a bar—and a complex network of pedestrian paths, gardens, and public spaces. Walls and floors of dwellings were movable so that students could design their own living spaces. Construction workers were offered design principles and constraints rather than finalized blueprints and were encouraged to generate sculptural forms. They did, and some returned on Sundays with their families to show off their art. The project's structural engineers were initially baffled by the level of spontaneity and playfulness, but they gradually proved capable of being reeducated into competent participants. All appears to have proceeded splendidly for some years, until the university administration became alarmed at its lack of control. With students away on vacation, they fired Kroll and halted further construction.[22]

Barrier-Free Design

During the past fifteen years, there has been substantial innovation in designing "barrier-free" equipment, buildings, and public spaces, which are responsive to the needs of physically disabled people. Much of the impetus comes from the disabled themselves, who have played roles ranging from mobilizing to redefine and assert their needs, to participating in inventing or evaluating design solutions.[23] For example, prototypes of the Kurzweil Reading Machine—which uses computer voice-synthesis to read typed text aloud—were tested by more than a hundred and fifty blind users. In an eighteen-month period these users made over a hundred recommendations, many of which were incorporated into later versions of the device.[24]

The movement for barrier-free design—now pervasively apparent in the profusion of ramps and modified restrooms in public places—is directly responsive to table 1, criterion B. It also represents an exciting prototype for the sort of activity embodied in my recommendation for a reconnaissance of needs, discussed earlier. Here we have nonmarket, democratic design criteria, often formulated and applied by disabled laypeople, being used to help define both individual and collective needs, including access to public space. Moreover, their evaluation is not of just one technology at a time—the norm in most technology assessments—but of our entire technological and architectural environment. The latter is a crucial distinction, because often the democratically significant aspects of design result from the simultaneous effects of many different kinds of technologies.

If next the range of democratic criteria were broadened (recall that criterion B is just one among a larger group that I suggest in table 1) and then the participants expanded beyond the disabled population, we would be moving toward a healthy capacity to generate a more democratic technological order.

Feminist Design

Engineering, architecture, and planning are today male-dominated professions. What would happen if women played a greater part in R&D and design? One indication originates from feminists, who have long been critical of housing designs and urban layouts that tend to enforce social isolation and arduous, unpaid domestic labor upon women. As more and more households deviate from the convention of two parents and children supported financially by a working husband, such criticism has sharpened.

One historical response has been to suggest that if women were more actively engaged in the design process, they might promote more shared neighborhood facilities (such as day care, laundries, or kitchens) and closer colocation of homes, workplaces, and commercial and public facilities. Realized examples already exist in London, Stockholm, and Providence, Rhode Island.[25] Another approach has been pioneered by artist and former overworked-mother Frances GABe, who has devoted thirty years to inventing a self-cleaning house. "In GABe's house, dishes are washed in the cupboard, clothes are cleaned in the closets, and the rest of the house sparkles after a humid misting and blow dry!"[26] Still other feminists seek alternatives to proletarianized female office work and to new reproductive technologies that could divorce women from control of their own bodies.[27]

Each of these feminist projects lends support to the basic contention that broadened opportunities to participate in R&D and design would indeed lead to a meaningfully broadened array of technological alternatives.

Third-World Design

There are numerous instructive cases of lay participation in designing technologies adapted to traditional or developing communities. For example, a Guatemalan appropriate technology station was approached in 1976 by village women who complained of excess wood consumption and smoke from their traditional, open-fire cooking methods. In response, researchers developed a clever, hardened-mud cookstove, called the "Lorena," that burns significantly less wood. Village women played a crucial role in evaluating prototypes—insisting, for example, that the original floor-level design be raised off the ground. The "final" design, now widely adopted, is made of inexpensive, locally available materials, and lends itself to user self-construction and redesign in conformance with local conditions.[28]

Another Third World example reveals a common sort of obstacle to participatory design. In the mid-1970s the remote Cree Indian community of Big Trout Lake, Canada, was struggling to cope with sewage-contaminated water supplies. In response, an agency of the Canadian government proposed an expensive sewage treatment plant that would have served only the small, non-Indian segment of the community. The Cree objected, taking the unusual step of hiring a participatory research facilitator and their own small team of technical consultants. Collaborating closely, the tribe and consultants developed an array of alternative, less-expensive waste treatment solutions that would be able to serve the entire community. Rather than welcome this sort of community involvement, the Canadian government reacted defensively and the project stalled.[29]

Designing with Children

In the 1960s city planner Mayer Spivack had a chance to help Boston children restore a neglected playground. Together, they designed playground equipment that could be built—and continuously redesigned and rebuilt—by kids themselves, using industrial surplus and other scavenged materials. The children quickly exhibited boundless exuberance in fashioning their own, ever-changing, public playspace. But local adults, who were not significantly involved in the process, objected to the youths' makeshift aesthetic and insisted that the city government demolish the playground.[30]

The story of American architect Bob Leathers is more hopeful. Since 1970 he has helped more than three hundred communities design, finance, and construct their own playgrounds, using a process partly derivative of an old-fashioned barn raising. Children—preschoolers and up—play a significant part in all phases, including design. Both the social process, and the new playgrounds themselves, have tended to engender a heightened experience of convivial, local community.[31]

POLITICAL IMPLICATIONS

The wealth of examples of participation in R&D and design allows one to make some tentative generalizations. First, it seems clear that the issue of laypeople's competence to participate meaningfully in R&D and design does not pose insurmountable problems. On the contrary, laypeople can certainly help identify social needs and concerns, and thus contribute to developing functional specifications for new technologies. In addition, their deep knowledge of local cultural and environmental circumstances can prove vital to the R&D process—as in developing the Lorena stove or Big Trout Lake's sewage treatment alternatives.[32] Finally, when a design process is effectively structured, laypeople can potentially become full partners in developing technical design specifications or design solutions, as in UTOPIA and London's technology networks.

Second, successful participatory design seems often to require, among other things, a conscious effort to cultivate mutual respect and trust among collaborating laypeople and technical experts. In some cases, formal "participation facilitators"—who need not necessarily be technically trained themselves—will be required. The Cree of Big Trout Lake hired such a facilitator. A supporting program of "education for participation" may also prove helpful, although its curriculum might need to focus as much on design politics and on nurturing self-confidence as on technical know-how. "Education for participation . . . is not education about aesthetics, or about cost-benefit . . ., it is education about power."[33]

Nor are laypeople the only ones with much to learn. As in the case of the structural engineers who worked on Brussels' Zone Sociale, technical experts may need assistance in learning how to learn both from and with laypeople. Moreover, technical "experts" are in fact *not* expert in the specific, paramount problem of designing democratic technologies. Indeed, both their training and professional interest may place them at some disadvantage, relative to nonexpert citizens, in being concerned or competent to seek democratic designs.[34]

Third, today the vast majority of R&D occurs within three institutional settings: corporations, government laboratories, and universities. In contrast, most examples of participatory design have occurred in some other setting—sometimes with assistance from iconoclasts employed within major institutions, and generally on the social periphery where financial and other resources are scarce.[35] Even the Swedish UTOPIA project had to struggle to receive government support, and then had to restrict itself to designing computer software, because the available funding was insufficient to undertake complementary hardware R&D. Indeed, as we have seen, many participatory design exercises have encountered outright opposition from powerful institutions. Hence it seems fair to suggest that the absence of more examples of participatory R&D has much less to do with issues of layperson competency than it does with economic, political, and cultural resistance.

It follows that if participatory design of democratic technologies is ever to become normal social practice, then political pressure must be developed for diverting economic resources to it, or else for incorporating it into the core social institutions that today undertake most R&D. Universities—or divisions of government research facilities that are not engaged in classified military research—may be the easiest places to start because they receive public subsidy and must serve the public interest. Participation by workers—rather than by outside citizens—in corporate R&D is also a realistic near-term target, and experiments with it are already becoming more widespread. In contrast, we can expect that insinuating general citizen participation into either military or corporate R&D will prove more challenging. Among other things, it would generally require some relaxation of the laws governing military and trade secrecy, or a corresponding strengthening of the laws governing freedom of information and expression.[36]

None of these suggested reforms are politically trivial. On the other hand, advocates of participatory R&D and design have often elected to state their case entirely in terms of the material interests of current nonparticipants. Or else they speak in terms of the contribution that participation can make to improve worker productivity or to more satisfactory design solutions. These can be reasonable arguments, and sometimes effective ones. But they neglect the specific moral argument that the opportunity to participate in R&D and design is a matter of moral right—essential to individual moral autonomy, human dignity, democratic self-governance, and the possibility of generating technologies that are themselves more compatible with democracy. Moral arguments, when they are compelling, are not only right to make because they are moral, but also because—in virtue of being moral—they can be effective. Throughout this century many groups that were otherwise disadvantaged have, after all, made significant strides by nonviolently asserting their human rights. In the technological domain, we already have a working precedent in the movement among disabled people for barrier-free design. Perhaps the challenge is to see whether those who are not disabled can learn something about conducting the politics of technology from those who are.

This extended discussion of just one of my recommendations—to focus on participation in R&D—is not definitive. But it should be sufficient to suggest the sort of elaboration that could be developed for the rest.

CONCLUSION

Building on the insight that technologies are freedom-conditioning social structures, I have suggested that it is both morally and democratically imperative that citizens become empowered to participate in all phases of technological politics, including evaluation, choice, and governance. I have dwelt especially on the little-heeded necessity of broadened participation in technological research, development, and design, complemented by the substantive need to ensure that our technologies become more compatible with establishing and reproducing democracy through time. In the language of my introduction, the "nuts and bolts of democracy" must become more than a metaphor.

In contrast with this vision, we can interpret contemporary technological politics in terms of several interrelated components. Many of the most important decisions are actually made via a covert politics that occurs within corporate headquarters and government bureaucracies; or via the tacit politics of the economic marketplace.[37] Meanwhile, the overt, publicly enacted politics of technology might be characterized as comprising, roughly, the wrong people posing the wrong questions of the wrong technologies at the wrong time.

Think, for example, of public controversies that have engulfed the proposed fluoridation of public water supplies, siting of nuclear power plants, basing of MX missiles, office automation, or new techniques in genetic engineering.[38] Time and again we witness powerful political actors—for example, professional politicians, government administrators, and corporate representatives, complemented by the activities of an unelected, politically unaccountable, and democratically insensitive technical elite[39] and glorified by sensationalist, superficial, and elite-dominated press coverage[40]—ritualistically debate the economic costs and consequences, military security implications, health and safety risks, or environmental impacts of the current crop of fashionably topical technological developments. The addressed questions are not inconsequential, but this system of public discourse and decision making nevertheless

- Excludes lay citizens from anything but a trivial role;
- Often raises questions publicly only long after many of the most important decisions—for example, over the available range of technological alternatives—have already been made elsewhere;
- Evaluates technologies almost exclusively on a case-by-case basis, thus discouraging the possibility of acting upon, or even identifying, socially significant complementarities and synergisms among ostensibly unrelated technological developments;[41]
- Focuses on just a few "sexy" or "cutting edge" technologies, to the virtual exclusion of the vast majority of emerging and— even more so—existing technologies;
- Directs attention primarily toward these publicized technologies' promised material, focal purposes, at the expense of often-more-important cultural and other nonfocal social consequences; and
- Leaves the question of the debated technologies' structural bearing upon the possibility of democracy as only the most important of many questions that are never asked.

So much for democratic technological procedure. Since the question of democracy is never systematically addressed, we might expect that at best contemporary technologies would prove in substantive terms to be a mixed lot: some favorable to democracy, some not, and the overall result a product of serendipity rather than human choice.

To some extent that expectation appears to be borne out. But if the design criteria listed in figure 1 are any indication, the balance within contemporary technological orders appears actually to be skewed in decidedly antidemocratic directions.

Might that result have something to do with the erosion of participatory cultures, and of social contexts for the development and expression of citizenship?[42] Or might it be related to modern tendencies toward loneliness, fragmentation of self, dissipation of purpose, narcissism, disempowerment, insecurity, and anomie?[43] Is it a result of the supplanting of moral and political autonomy by relative automatism?[44] Does it bear out the observation that in contemporary politics, "Voters elect people to be citizens. . . . But the voters are not citizens, just voters"?[45]

I suspect that the answer is yes. This is troubling in itself and also implies that existing technologies will provide some hindrance against future democratization efforts—both those that are technologically oriented and more generally.

It will not always be easy, but we can and must do better. I thus uphold a vision of aspiring citizens struggling to transform the technological order from an arbitrary or antidemocratic social force into a substantive constituent and expression of human freedom.

NOTES

I am indebted to Marcie Abramson, Brian Baker, Michael Black, Iain Boal, Larry Bucciarelli, Vary T. Coates, Paul T. Durbin, David Elliott, William W. Hogan, Todd R. LaPorte, Laura Nader, Richard B. Norgaard, Wolf Schäfer, and Jeff Scheuer for their many helpful comments on earlier drafts of this essay. My research was made possible, in part, through the award of a Ciriacy-Wantrup Postdoctoral Fellowship from the University of California, Berkeley (1986–87), and through the generosity of the Menemsha Fund and the Institute for Resource and Security Studies. None of the preceding persons or organizations should, however, be construed as necessarily endorsing the chapter's views.

1. See, for example, Lester C. Thurow, "A World-Class Economy: Getting Back into the Ring," *Technology Review* 88, no. 6 (August/September 1985): 26–37; Ralph Landau and Nathan Rosenberg, eds., *The Positive Sum Strategy: Harnessing Technology for Economic Growth* (Washington, D.C.: National Academy Press, 1986).

2. See also Langdon Winner, *The Whale and the Reactor: A Search for Limits in an Age of High Technology* (Chicago: University of Chicago Press, 1986), p. 10.

3. Richard E. Sclove, "Technology and Freedom: A Prescriptive Theory of Technological Design and Practice in Democratic Societies" (Ph.D. diss., Massachusetts Institute of Technology, 1986).

4. See, for example, Sclove, "Technology and Freedom," pp. 309–419; and Winner, "Do Artifacts Have Politics," in *The Whale and the Reactor*, pp. 19–39.

5. On technologies as contingent social products see, for example, Arnold Pacey, *The Maze of Ingenuity: Ideas and Idealism in the Development of Technology* (Cambridge: MIT Press, 1976); David F. Noble, *Forces of Production: A Social History of Machine Tool Automation* (New York: Alfred A. Knopf, 1984); Michael J. Piore and Charles F. Sabel, *The Second Industrial Divide* (New York: Basic Books, 1984); Ruth Schwartz Cowan, "The Roads Not Taken: Alternative Social and Technical Approaches to Housework," in *More Work for Mother: The Ironies of Household Technology from the Open Hearth to the Microwave* (New York: Basic Books, 1983), pp. 102–50.

On social structures influencing but not determining social experience see, for example, Marshall Sahlins, *Culture and Practical Reason* (Chicago: University of Chicago Press, 1976); Roberto Mangabeira Unger, *Social Theory: Its Situation and Its Task* (Cambridge: Cambridge University Press, 1987).

6. For the philosophical grounding of this and subsequent democratic theory see Sclove, "Technology and Freedom" (note 3, above), chapter 2.

7. For further suggestions—not all of which I endorse, but which I consider eminently worthy of debate—see Benjamin Barber, *Strong Democracy: Participatory Politics for a New Age* (Berkeley: University of California Press, 1984),

pp. 261–31; Samuel Bowles, David M. Gordon, and Thomas E. Weisskopf, *Beyond the Wasteland: A Democratic Alternative to Economic Decline* (Garden City, N.Y.: Doubleday, 1983); Michael Goldhaber, *Reinventing Technology: Policies for Democratic Values* (New York: Routledge & Kegan Paul, 1986); Carmen J. Sirianni, "Production and Power in a Classless Society: A Critical Analysis of the Utopian Dimensions of Marxist Theory," *Socialist Review* 11, no. 5 (September–October 1981): 33–82; International Association of Machinists and Aerospace Workers, *Let's Rebuild America* (Washington, D.C.: Kelly Press, 1984), especially the proposed "Technology Bill of Rights."

8. For examples—some specifically technological in orientation, some not—see Barry M. Casper and Paul David Wellstone, *Powerline: The First Battle of America's Energy War* (Amherst; University of Massachusetts Press, 1981); Claire Huchet Bishop, *All Things Common* (New York: Harper & Row, 1950); Harry C. Boyte, *The Backyard Revolution: Understanding the New Citizen Movement* (Philadelphia; Temple University Press, 1980); Lawrence Goodwyn, *The Populist Moment: A Short History of the Agrarian Revolt in America* (Oxford: Oxford University Press, 1978); Philip Hallie, *Lest Innocent Blood Be Shed: The Story of the Village of Le Chambon and How Goodness Happened There* (New York: Harper & Row, 1980); Mohandas K. Gandhi, *Non-Violent Resistance* (New York: Schocken, 1983); David J. Garrow, *Bearing the Cross: Martin Luther King, Jr., and the Southern Christian Leadership Conference* (New York: William Morrow, 1986).

9. See, for example, Walter C. Patterson, "The Open University Tackles Control of Technology," *Bulletin of the Atomic Scientists,* March 1980, pp. 56–57.

10. See, for example, Tom Schlesinger, John Gaventa, and Juliet Merrifield, *How to Research Your Local Military Contractor* (New Market, Tenn.: Highlander Center, 1983).

11. For an address list of community design centers in the United States, see C. Richard Hatch, ed., *The Scope of Social Architecture* (New York: Van Nostrand Reinhold, 1984), pp. 346–47.

12. See, for example, Robert Gilman, "Four Steps to Self-Reliance: The Story behind Rocky Mountain Institute's Economic Renewal Project," *In Context,* no. 14 (Autumn 1986): 41–46.

13. See Todd R. LaPorte, "Technology as Social Organization: Implications for Policy Analysis," IGS Working Papers 94-1, Institute of Governmental Studies, University of California, Berkeley, January 1984.

14. See, for example, the chapter on public participation in U.S. Congress, Office of Technology Assessment, *Coastal Effects of Offshore Energy Systems: An Assessment of Oil and Gas Systems, Deepwater Ports, and Nuclear Powerplants Off the Coast of New Jersey and Delaware* (Washington, D.C.: Government Printing Office, 1976), pp. 255–79.

15. A great deal of theoretical and practical work has been done in support of corporate democratization. See, for example, Bowles, Gordon, and Weisskopf, *Beyond the Wasteland*; Robert A. Dahl, *A Preface to Economic Democracy* (Berkeley: University of California Press, 1985); David W. Ewing, *Freedom Inside the Organization: Bringing Civil Liberties to the Workplace* (New York: McGraw-Hill, 1977).

16. See, for example, Garry Wills, "Power Unchecked," *The Washington Post Magazine,* 28 June 1987, pp. 30–39; Carl Barus, "Military Influence on the Electrical Engineering Curriculum Since World War II," *IEEE Technology and Society Magazine* 6, no. 2 (June 1987): 3–8; David F. Noble, "Command Perform-

ance: A Perspective on Military Enterprise and Technological Change," in *Military Enterprise and Technological Change: Perspectives on the American Experience*, ed. M. Smith (Cambridge: MIT Press, 1985), pp. 329–46; Suzanne Gordon and Dave McFadden, eds., *Economic Conversion: Revitalizing America's Economy* (Cambridge: Ballinger, 1984); Paul Kennedy, *The Rise and Fall of the Great Powers: Economic Change and Military Conflict from 1500 to 2000* (New York: Random House, 1987).

17. As per my earlier arguments regarding the moral primacy of democracy over narrowly economic considerations, it does not really matter whether or not vast corporate enterprises are economically more "efficient" than smaller ones. But there is evidence to suggest that in any case they are not. See Walter Adams and James W. Brock, *The Bigness Complex* (New York: Pantheon, 1986).

18. My sources of information on Lucas include Mike Cooley's *Architect or Bee? The Human/Technology Relationship* (Boston: South End Press, 1982); and "The Taylorisation of Intellectual Work," in *Science, Technology and the Labour Process: Marxist Studies,* ed. L. Levidow and B. Young, (London: CSE Books, 1981), 1:46-65; Hilary Wainwright and David Elliott, *The Lucas Plan: A New Trade Unionism in the Making?* (London: Allison and Busby, 1982); Walter C. Patterson, "The Open University Tackles Control of Technology," *Bulletin of the Atomic Scientists,* March 1980, pp. 56–57; and David Elliott, interview with author, December 1984.

19. Mike Cooley, interviews with author, December 1984 and December 1986; David Elliott, "The GLC's Innovation and Employment Initiatives," TPG Occasional Paper 7 (Milton Keynes: Technology Policy Group of the Open University, April 1984); Veronica Mole and Dave Elliott, "The GLEB Innovation Experiment," in *Enterprising Innovation: An Alternative Approach* (London: Frances Pinter, 1987), pp. 72–104.

20. Robert Howard, "UTOPIA: Where Workers Craft New Technology," *Technology Review,* April 1985, pp. 43–49; Andrew Martin, "Unions, the Quality of Work, and Technological Change in Sweden," in *Worker Participation and the Politics of Reform,* ed. C. Sirianni (Philadelphia: Temple University Press, 1987), pp. 95–139.

21. See, for example, Christopher Alexander et al., *The Production of Houses* (Oxford: University Press, 1982); Hatch, *The Scope of Social Architecture*; Robert Somer, *Social Design: Creating Buildings with People in Mind* (Englewood Cliffs, N.J.: Prentice-Hall, 1983), pp. 110–29.

22. Lucien Kroll, "Anarchitecture," in Hatch, *The Scope of Social Architecture,* pp. 166–85.

23. See, for example, Virginia W. Stern and Martha Ross Redden, eds., *Technology for Independent Living* (Washington, D.C.: American Association for the Advancement of Science, 1982); Gary O. Robinette, ed., *Barrier-Free Exterior Design: Anyone Can Go Anywhere* (New York: Van Nostrand Reinhold, 1985); Patricia Leigh Brown, "Designs Take Heed of Human Frailty," *New York Times,* 14 April 1988.

24. See two articles in Stern and Redden, *Technology for Independent Living:* Michael Hingson, "The Consumer Testing Project for the Kurzweil Reading Machine for the Blind," pp. 89–90, and Raymond Kurzweil, "The Development of the Kurzweil Reading Machine," pp. 94–96.

25. Dolores Hayden, *Redesigning the American Dream: The Future of Housing, Work, and Family Life* (New York: Norton, 1984), pp. 163–70.

26. Jan Zimmerman, *Once upon the Future: A Woman's Guide to Tomorrow's*

Technology (New York: Pandora, 1986), pp. 36–37. See also Frances GABe, "The GABe Self-Cleaning House," in *The Technological Woman: Interfacing with Tomorrow,* ed. J. Zimmerman (New York: Praeger, 1983), pp. 75–82.

27. See, for example, Joan Rothschild, ed., *Machina ex Dea: Feminist Perspectives on Technology* (New York: Pergamon, 1983); Wendy Faulkner and Erik Arnold, eds., *Smothered by Invention: Technology in Women's Lives* (London: Pluto, 1985).

28. Donald Wharton, "Designing with Users: Developing the Lorena Stove, Guatemala," in *Experiences in Appropriate Technology,* ed. R. Mitchell (Ottawa: Canadian Hunger Foundation, 1980), pp. 21–26.

29. "Users Making Choices in a Fragile Environment, Canada," in Mitchell, *Experiences in Appropriate Technology,* pp. 47–58.

30. Mayer Spivack, "The Political Collapse of a Playground," *Landscape Architecture* 59, no. 4 (July 1969): 288–92.

31. Richard Wolkomir, "A Playful Designer Who Believes that Kids Know Best," *Smithsonian,* August 1985, pp. 106–15.

32. Compare D. J. Gamble, "The Berger Inquiry: An Impact Assessment Process," *Science* 199 (3 March 1978): 946–52; Sheldon Krimsky, "Epistemic Considerations on the Value of Folk-Wisdom in Science and Technology," *Policy Studies Review* 3, no. 2 (February 1984): 246–62; Jayanta Bandyopadhyay and Vandana Shiva, "The Legitimacy of People's Participation in the Formulation of Science and Technology Policy: Some Lessons from the Indian Experience," in *Citizen Participation in Science Policy,* ed. J. Petersen (Amherst: University of Massachusetts Press, 1984), pp. 96–106.

33. Colin Ward, *Housing: An Anarchist Approach,* 2d ed. (London: Freedom Press, 1983), p. 119. Ward was commenting on participation in urban planning, but his insights have broader applicability. See also Robert Sommer, *Social Design,* pp. 119–22.

34. There is certainly evidence of mid-level managers, many of whom are scientists or engineers, resisting technologies that facilitate workplace democratization—even when labor productivity has demonstrably improved. See, for example, Daniel Goleman, "Why Managers Resist Machines," *New York Times,* 7 February 1988.

35. See also Pam Linn, "Socially Useful Production," *Science as Culture,* no. 1 (1987): 105–38.

36. See Sissela Bok, *Secrets: On the Ethics of Concealment and Revelation* (New York: Pantheon, 1982), especially chapters 10–14; Goldhaber, *Reinventing Technology,* especially chapter 10; Ewing, *Freedom Inside the Organization,* especially chapter 6; John Shattuck and Muriel Morisey Spence, "The Dangers of Information Control," *Technology Review* 91, no. 3 (April 1988): 62–73.

37. See, for example, Bruno Latour, *Science in Action: How to Follow Scientists and Engineers through Society* (Cambridge: Harvard University Press, 1987); Wiebe E. Bijker, Thomas P. Hughes, and Trevor J. Pinch, eds., *The Social Construction of Technological Systems* (Cambridge: MIT Press, 1987); Nathan Rosenberg, *Inside the Black Box: Technology and Economics* (Cambridge: Cambridge University Press, 1982); Noble, *Forces of Production*; Cowan, *More Work for Mother.*

38. See, for example, Allan C. Mazur, *The Dynamics of Technological Controversy* (Washington, D.C.: Communications Press, 1981); Dorothy Nelkin, ed., *Controversy: Politics of Technical Decisions,* 2d ed. (Beverly Hills: Sage, 1984).

39. See, for example, Sclove, "Energy Policy and Democratic Theory," in

Uncertain Power: The Struggle for a National Energy Policy, ed. D. Zinberg (New York: Pergamon, 1983), pp. 41–48; Sclove, "Technology and Freedom," pp. 76–81; Michael Mulkay, "The Mediating Role of the Scientific Elite," *Social Studies of Science* 6, nos. 3–4 (September 1976): 445–70; Brian Wynne, "The Rhetoric of Consensus Politics: A Critical Review of Technology Assessment," *Research Policy* 4 (1975): 108–58; David Dickson and David Noble, "By Force of Reason: The Politics of Science and Technology Policy," in *The Hidden Election: Politics and Economics in the 1980 Presidential Campaign,* ed. T. Ferguson and J. Rogers (New York: Pantheon, 1981), pp. 260–312; Paul K. Feyerabend, *Science in a Free Society* (London: NLB, 1978).

40. Dorothy Nelkin, *Science, Technology and the Press* (New York: Freeman, 1987).

41. For historical examples of such synergisms see, for example, Michel Foucault, *Power/Knowledge: Selected Interviews and Other Writings, 1972–77,* ed. Colin Gordon (New York: Pantheon, 1980); Paul Rabinow, ed., *The Foucault Reader* (New York: Pantheon, 1984); Albert Borgmann, *Technology and the Character of Contemporary Life* (Chicago: University of Chicago Press, 1984); Cowan, *More Work for Mother;* Sclove, "Technology and Freedom," pp. 96, 388–91, chapter 5 (especially 510–18), and 788–89.

42. See John Dewey, *The Public and Its Problems* (1972; reprint, Chicago: Swallow Press, 1954); Robert S. Lynd, Foreword to *Business as a System of Power,* by Robert A. Brady (New York: Columbia University Press, 1943), pp. vii–xviii; Goodwyn, *The Populist Movement.*

43. For documentation and discussion of these ills, see, for example, Theodore Roszak, *Where the Wasteland Ends: Politics and Transcendence in Post-Industrial Society* (Garden City, N.Y.: Doubleday, 1973); Roberto Mangabeira Unger, *Knowledge and Politics* (New York: Free Press, 1975); Robert J. Lifton, *The Broken Connection: On Death and the Continuity of Life* (New York: Touchstone, 1980); Erazim Kohak, *The Embers and the Stars: A Philosophical Inquiry into the Moral Sense of Nature* (Chicago: University of Chicago Press, 1984); Borgmann, *Technology and the Character of Contemporary Life*; Robert N. Bellah, et al., *Habits of the Heart* (Berkeley: University of California Press, 1985).

44. See, for example, Lawrence Kohlberg, *The Philosophy of Moral Development: Moral Stages and the Idea of Justice* (San Francisco: Harper and Row, 1981), pp. xxxiii, 88, 237, 242; Robert Kegan, *The Evolving Self: Problem and Process in Human Development* (Cambridge: Harvard University Press, 1982), pp. 207–15, 243–49 (Kegan writes as one who is sympathetic to criticisms that Kohlberg's research may be sexually or culturally biased); Dorothy Lee, *Freedom and Culture* (Englewood Cliffs, N.J.: Prentice-Hall, 1959); Jacob Needleman, *Consciousness and Tradition* (New York: Crossroad, 1982), pp. 72–87.

45. Karl Hess, "The Politics of Place," *Coevolution Quarterly,* no. 30 (Summer 1981), p. 6.

Part 5
Engineering Education

Bridging Gaps in Philosophy and Engineering

TAFT H. BROOME, Jr.

INTRODUCTION

In this essay, I intend to approach the subject of philosophy and engineering from three vantage points. To say that the subjects of my discussion are philosophy *and* engineering would mean that I treat them as two distinct learned disciplines, norms of professional practice, and systems of shared values—that is, as two distinct cultures. Many examples of such an approach can be found in ethics literature, in which engineers' professional situations are used as starting points for studying their moral behaviors. The well-known Baum and Flores texts are examples.[1]

To say that the subject of this paper is philosophy *in* engineering would mean that I preserve the distinctness of the two learned disciplines but use the formal tools of philosophical inquiry to examine some aspect of engineering.[2] Petroski's book, *To Engineer Is Human*[3] is one example; the author examines the role of failure in the evolution of engineering paradigms.

Finally, I will consider a third mode of discussion, philosophy *of* engineering, in which the problems subjected to examination by philosophy *in* engineering are synthesized into a coherent theory. I have argued before for one example of this mode of discussion[4] and will consider it again here.

The three modes in which philosophy engages engineering all meet with a resistance that is rooted in the cultures of philosophy and engineering. The purpose of this essay is to inquire into the nature of this resistance.

My objectives are to (1) describe a sample of the values residing in the cultures of philosophy and engineering; (2) argue that these values resist philosophical engagement of engineering; and (3) suggest means of suppressing or compensating for this resistance.

Thus, the gaps mentioned in the title do not refer to any hiatus between philosophy and engineering but to inadequacies in each of them that frustate attempts to consummate their engagement. "Bridging gaps" therefore refers to successful efforts to compensate for these inadequacies. My thesis is that bridges over the gaps on either side will be the essentials of an emerging culture that I label *science, technology, and human values* (STHV).

My argument rests on three fundamental assumptions. First, gaps within the cultures of philosophy and engineering are essentials of these cultures. Thus, there is no need to suggest how either the culture of philosophy or the culture of engineering should be altered. Second, the emerging STHV culture is evolving its own distinct system of shared values; although the system's component values are derived mainly from the cultures of philosophy, engineering, and history (among others), the system's logic—its basis for connecting these values and resolving conflicts among them—will be unique. The aim of this paper is to show how this new STHV culture can bridge the gaps within philosophy and engineering. The third assumption is that the rationale for the choice of values within this system, and for the form of its logic, will be grounded in ideas obtained from or that help consummate what I call *philosophical engagements of engineering*. But this leaves a question: How can STHV bridge these gaps when its value system is assumed to be built up from bridges over them?

BACKGROUND: STAGNATION OF STHV

Philosophical engagements of engineering form causal links between ideas in science and products of technology. As we shall see, these links are needed to explain, for instance, why engineers intentionally introduce hazards into nuclear technologies in excess of those hazards normally associated with the uses of science; or how engineering provides engineers reasons to persist in developing the SDI technology in the face of substantial opposition from scientists. Because STHV can not prosper without these engagements, the presence of gaps in the cultures of philosophy and engineering indicates stagnation in STHV.

Most of the bridges laid over these gaps have the collaborative works of philosophers and engineers in the adolescent field of philosophy *and* engineering—particularly the field of engineering ethics. Productivity in the infantile field of philosophy *in* engineering has apparently been the result of efforts of engineers who have pursued philosophical studies more than casually in order to publish for an engineering readership. Koen's work, *Toward a Definition of the Engineering Method*[5] is a noteworthy example. Productivity in the embryonic field of philosophy *of* engineering has resulted from engineers having one of two missions: either to transmit the essences of engineering to nonengineering undergraduate students[6] or to inquire into the nature of engineering mainly for the benefit of philosophers.[7]

Progress in development of the first mode of discussion has all but halted: no new ideas are emerging from case studies, and further development of discussions in the second mode are needed to provide the kinds of explanations mentioned above. Progress in development of the second and third modes of discussion is extremely slow because of manpower shortages.

Getting more engineers to pursue serious study in philosophy is a difficult task because the desired career rewards are often not seen as forthcoming. Engineering faculty promotion committees give virtually no weight to publications in philosophy, and there is little evidence elsewhere to indicate that the engineer/philosopher career path is a profitable one. Getting philosophers to pursue serious study in engineering is a difficult task also because of the time investment required to navigate the prerequisite structures of the engineering curriculum. Thus, STHV will continue to stagnate unless bold new initiatives are mounted to recruit engineer/philosophers to bridge the gaps in the cultures of philosophy and engineering.

In anticipation of such bold new initiatives, some of these gaps will be described and means of bridging them will be offered.

PRECISION AND RIGOR

Precision and rigor may at first seem unlikely candidates for sources of resistance to any kind of scholarly endeavor. But to anyone who has endured an interdepartmental faculty meeting, the scholarly commitment to highly precise and rigorous statements can be a formidable barrier to progress—particularly when crossing disciplinary boundaries. The knowledge base shared by the participants may be so weak as to prevent them from formulating statements generally regarded as meaningful. Such meetings are known to become so frustrating that their participants find themselves embroiled in a general unwillingness to agree upon such lower levels of precision and rigor as may be compatible with the intended use of these statements. Nevertheless, highly precise and rigorous interdisciplinary statements are produced by some groups. What, then, are the keys of productivity for such groups?

My response to this question results from twelve years of active experience as a participant/observer (or in some cases, engineer/ philosopher) in various productive groups, and in some unproductive ones, each consisting of humanists and technologists. Two of the productive groups consisted of philosophers and engineers headquartered at Rennselaer Polytechnic Institute and at Illinois Institute of Technology. The other productive group consisted mainly of philosophers and physicians at the Kennedy Institute for Bioethics at Georgetown University. The unproductive ones were short-lived task forces organized by certain universities and professional societies.

Common to these groups was the consensus among their participants that the causes of their groups were worthy. Common to the productive groups and absent from the others were the affirmations of each participant that something of high value was at stake in these causes. Evidence of what was personally at stake has emerged mostly in the form of individual achievements: at least four of the participants founded three scholarly journals; books and/or scholarly papers were published by almost everybody; at least five participants assumed leadership of ethics programs in engineering and scientific professional societies; and two STHV programs were established largely because of the efforts of these participants. Perhaps a clue to the whereabouts of our key lies in the following observation: In the unproductive groups, considerable attention was given to what was *wrong* about something someone said. In the productive groups, most of the time was spent exploiting what was *right* about something someone said.

This observation would suggest that the key lies in the attitude that, in their early stages of development, interdisciplinary studies can benefit more from the value of exploiting the right in something than from finding error in it. To the extent that the converse is true in established cultures, error-finding as a value can be said to impede progress in most philosophical attempts to engage engineering. On the other side, the former value tends to suppress the effect of or compensate for this resistance. To say the same thing, until STHV matures to the point where it can progress without worrying about scholarship in impeccable condition, precision and rigor will be gaps in the cultures of philosophy and engineering; and tolerance of imprecision will be STHV's bridge over these gaps.

Consider Koen's mode of discussion on engineering heuristics.[8]

This discussion might not have taken place if philosophers had the veto, because not all of the tools of philosophical inquiry were implemented. Nevertheless, Koen's work presents ideas that are indispensable parts of a rigorous third-mode discussion. Thus, to produce more ideas indispensable to philosophical engagement of engineering, STHV might well benefit from philosophical papers, research projects, or other works of engineers that may not find acceptance in established scholarly cultures. As we shall see, ample precedents exist for anticipating that when STHV becomes an established culture, it will not seek its credibility from other established cultures.

But what is at stake? Since tolerance of imprecision can not be allowed to run unchecked, are STHV scholars sufficiently motivated to expend the energy required to mine the depths of two heretofore disparate disciplines for their precious essences?

Clearly, the long-term solution to the problem of bridging the precision and rigor gaps in philosophy and engineering requires the production of scholars who share a substantial knowledge base spanning these two cultures. For now, the argument that STHV has begun to stagnate and will likely be dismantled if it can not bridge these gaps will have to do.

CONTEMPLATING AND DOING

The philosopher Peter Caws has observed:

> The opposition of theory and practice, and the scorn of the latter, . . . has ancient roots, but we have perhaps not understood them. It was certainly not, as is sometimes supposed, a simple question of slavery and nobility; rather it was perceived that there are two different and independent manners of relating to the world. Thus Aristotle in the *Nicomachean Ethics* says, "For a carpenter and a geometer investigate the right angle in different ways; the former does so in so far as the right angle is useful for this work, while the latter inquires what it is or what sort of thing it is; for he is a spectator of the truth." . . . There is no evidence that the Greeks despised carpenters. There is plenty of evidence that they admired geometers and indeed mathematicians in general, but the reason for this can easily be traced back to religious motives. Pure contemplation (according to) Aristotle is a function of the divine. . . .[9]

The engineer Samuel Florman, motivated by a similar observation, attributed an "anti-doing" bias to philosophers on account of their inclinations for writing:

Plato and Aristotle have the last word because they wrote down what they said. There is a nasty, unfair epigram which says, "Those who can, do, Those who cannot, teach." It could more accurately be paraphrased, "Many of those who can, do." Some of those who don't, write, and tend to criticize doing.[10]

Surely many engineers would join Florman in his view that meaningful responses from the engineering community to these criticisms require of engineers greater literary inclinations than are normally credited to them. The responses engineers would write, if inclined to do so, would likely reveal substantial respect for a contemplation/doing continuum, but general scorn for contemplation without doing. Moreover, in academic advising sessions, engineering faculty and their students can be observed treating core requirements in the humanities as nuisances. Philosophy, for example, is too often judged a triviality because engineering students are known to perform well in its introductory courses without exertion and because these courses are generally regarded as doubtful avenues to a good job.

The claim that contemplation is superior in scholarly value to doing has credibility wherever institutions of higher learning admit of a prime commitment to truth. This claim has raised questions within the academy that are highly charged politically. For example, the question, "Are humanists and other seekers of truth better fitted than professional school faculty to decide the fate of the university?" has long contributed to the divisiveness that has been sensed between philosophers and engineers on many campuses. Today, in view of shifting patterns in student career choices and federal research spending, challenges for humanists to attach more practical value to their work introduce divisive questions into the academy: "Has the engineer's time come at last?" and "Are engineers now better fitted than humanists to decide the fate of the university?" Questions of this sort haunt many campuses.

Political divisiveness, rooted in value conflicts between contemplation and doing, tends to resist formation of the kinds of collaborations between philosophers and engineers that can lead to greater productivity in all three modes—in philosophy *and, in* and *of* engineering. Thus, gaps exist in the two cultures where contemplation and doing reside. For STHV to bridge these gaps, its new value system must accommodate both contemplation and doing with equal status. Then, STHV's greatest asset to the academy may well prove to be its gift of enlightened values. But what opportunities exist, if any, for STHV scholars to influence the values of contemplation and doing held by their non-STHV colleagues and students?

I have argued that engineering should be part of everyone's education.[11] If STHV scholars were to carry this argument into their core curriculum design discussions, healthy debates might likely ensue. These debates would be forums in which a wide variety of non-STHV scholars could witness the use of theoretical frameworks in which contemplation and doing have equal status. These frameworks would then be at the disposal of a broad population of scholars for their use in evaluating their leadership and shaping policy in the academy.

AGREEMENT AND DISAGREEMENT

Engineering education is fraught with instances of the phenomenon commonly known as *the* right answer. Moreover, the right answer is always arrived at through proper implementation of *the* right procedure. In practice, individual judgment plays a greater role than in the classroom, but codes and standards remain dominant factors in the daily life of the engineer. While creative use of these factors separates good engineers from others, engineering can still be criticized for not emphasizing intellectual liberation and freedom—goals of the philosopher. Nevertheless, agreements on what the right answer is, and how to get it, are essential to the preparation of students for teamwork and leadership in situations requiring the orchestration of many persons and machines, and the application of sophisticated principles to unique, complex, life-threatening, economically and politically sensitive, legally precipitous, time-constrained problems. Successes in solving such problems produce in engineers Florman's "existential pleasures" and reinforce for them the value of an engineering education. Small wonder, then, that engineers are amazed each time William James is proved correct in his assertion, "There is only one thing that a philosopher can be relied on to do, and this is to contradict other philosophers."[12] No wonder at all, then, that engineers criticize philosophers for not finding *the* correct solution to every moral problem, saying, "If philosophers could find it, they would agree on it." Nor is it any wonder that engineers criticize philosophy for not equipping philosophers with *the* correct procedure for arriving at *the* correct solution.

Responses to these criticisms from the philosophical community may be summarized in Rollo May's observation: "The goal of perfectibility is a bastardized concept smuggled into ethics from technology, and results from a confusion between the two."[13] Philosophers know that multiple, publicly defensible arguments offering various solutions to many ethical and other philosophical problems can be found. They know that only the intellectually liberated and academically free can contemplate these arguments.

Philosophers characteristically have not been in need of interdisciplinary collaboration in order to facilitate their contemplations—history has amply demonstrated this point. But today philosophers in STHV studies characteristically have need of such collaborations with engineers.

The value of agreement in engineering resists productive use of the value of disagreement in philosophy, and vice versa. Philosophers often cannot communicate the validity of a given theoretical framework in the midst of theoretical diversity to engineers, and engineers often cannot communicate the validity of established engineering paradigms in the midst of divergent but equally defensible ideas to philosophers. Thus the engineer may expect little help from philosophers in his or her attempt to fashion *the* valid code of ethics for engineers, and the philosopher may expect little help from engineers in his or her quest for valid contrary points of view on a technological matter.

Such expectations of helplessness can be said to resist production of scholarship in philosophy *and* and *in* engineering. Gaps, therefore, can be said to exist in the cultures of philosophy and engineering, where values of agreement and disagreement reside. I would suggest that, to bridge these gaps, the emerging STHV culture needs to breed new scholars who are at home in both cultures.

As observed earlier, the bulk of productivity in philosophy of engineering has been contributed by engineers who have pursued philosophical studies seriously. These scholars have little need of help from others to assess the validity of their arguments. They feel comfortable constructing one theoretical framework for these arguments, and assessing its defensibility and usefulness with respect to other valid frameworks. They feel comfortable, also, in assessing the adaptability of established engineering paradigms to novel situations and the hazards of introducing new ideas into these situations. They know that the logical integrity of their arguments is *sometimes* better assured by solitary thinkers knowledgeable in both philosophy and engineering than by "committees" of philosophers and engineers. Ultimately, I anticipate STHV scholars will be bred possessing sufficient knowledge of both engineering and philosophy to advance studies in philosophy *and, in,* and *of* engineering without being separately either philosophers or engineers.

KNOWLEDGE AND WISDOM

Epistemology means theories of knowledge. "What is knowlege?" and "How do we get it?" are questions epistemology seeks to answer.

The sciences can be considered theories for claiming knowledge of things: in astronomy there is a theory of the universe in the aggregate; in physics, a theory about the constituents of the universe; in economics, a theory of the production and distribution of wealth; in psychology, a theory of the mind and its relation to the body; and so forth. The Kotarbinski school of praxiology defines engineering, too, as a science—the science of efficient action.[14] Other views also declare engineering to be a science of various sorts.[15]

However, engineering cannot be a science. Engineering projects are not necessarily designed to produce knowledge. First, these projects are often designed without controls; thus, hypotheses about cause-and-effect relationships are often difficult to prove. Second, the knowledge gained from such projects is only partially transferable to other situations because engineering systems are so complex that each can be regarded as unique and is thereby an experiment in and of itself. Finally, as Ronald Munson argued for medicine,[16] engineering is autonomous, in the sense that the credibility of its principles and methods for engineers does not require corroboration from the sciences.

An example: On the one hand, most scientists consider the meteor theory explaining the great extinctions to be more credible than competing theories because it corroborates better than others knowledge from many sciences—for example, astronomy, paleontology, and meteorology. On the other hand, scientists consider the SDI system to be infeasible because the system can not be tested except under full operating conditions. Engineers, however, make judgments all the time based on predictions about the performance of their works, about the feasibilities of these works in the absence of full-scale scientific tests; for instance, Krohn and Weingart report judgments of this kind in assessing the safety of nuclear power plants.[17] Martin and Schinzinger make similar observations and suggest more generally that engineering could be described as an experiment using the public as human subjects.[18] Engineering, then, does not rely on science to establish its legitimacy; it therefore cannot be considered a science.

Engineers are known to persist with the science conception of engineering while holding this science-independent legitimacy argument. Why?

I have argued that engineers use the science conception to gain public sanction of their business interests.[19] I have also argued that engineers abandon this view to explain risks associated with their work that exceed those normally accepted as inherent in science. Thus, the value of scientific "knowing" for business interests in the culture of engineering can be said to resist complete abandonment of an incorrect "scientific" conception of engineering.

Similar arguments could be made about an "applied science" conception of engineering. Applied science can be taken to be a theory about the use of scientific knowledge to achieve some practical end. For example, if we know that acids and bases combine to form salt and water, then the introduction of a base into an upset stomach caused by too much acid may produce relief. This is what applied science is all about—a practical problem is solved using a methodology consisting of scientific principles alone. There is error, however, in the applied-science conception because engineering methodology consists not of scientific principles alone but of scientific principles together with nonscientific practices. Moreover, the observation that engineering practices are often established and successfully used in advance of their scientific proofs suggests that the same arguments used to define engineering as applied science could be used—equally incorrectly—to define science as some sort of "redundancy" engineering.

Despite these arguments, philosophers tend to hold the applied-science view of engineering. Such a view allows them to argue, for example, that engineering has no legitimate place in the core curriculum since its cognitive content can be reduced to physics, chemistry, or another credible science. Peter Caws attributes the tenacity of the applied-science view to an elitist tradition of values shared by both philosophers and scientists: "Technology has continued for the most part to be relegated to the kitchen, and even philosophers who are beginning to pay attention to it do not always manage to avoid the patronizing tone of those who, to their surprise, have found hidden talent below the stairs. The assumption, all too readily made and acepted, that technology is to be defined as the practical application of scientific theory is symptomatic of this."[21] This elitist tradition was originated by Aristotle and handed down to philosophers by him. The maturation of philosophy of science extended the tradition to scientists. Thus, this elitist tradition of values in the culture of philosophy—reinforced by scientists—tends to uphold the incorrect applied-science conception of engineering.

The value of scientific knowing to business interests in the culture of engineering, and these elitist values in the culture of philosophy, are gaps in philosophy and engineering. By promoting incorrect conceptions of engineering, these values tend to resist correct philosophical engagements of engineering. To bridge these gaps, STHV must include in its system of values new correct conceptions of engineering. The basic ideas forming the embryonic stage of one such conception are outlined below.[21]

For Plato and Aristotle, Aquinas, Spinoza, and a host of others, wisdom has been the subject of considerable contemplation. To suit present purposes, wisdom will be defined as the ability to make the best use of knowledge, some tradition of practice, personal experience, and intuition to achieve some end.

Engineering can be considered as a set of theories claiming the wisdom to achieve purposive changes of things: in civil engineering there are theories for changing the land; in mechanical engineering, theories for changing heat energy into motion; in electrical engineering, theories for changing magnetic and other phenomena into electricity and back; and so forth. To judge which use of knowledge is the best use, the "engineer's imperative" is acknowledged: Make these changes subject to constraints of time, economics, and safety, among other things. Thus, the application of science alone to these problems of change is seldom possible. While more science in the methodology of engineering is valued more highly than less, observance of the engineer's imperative always requires insinuation of a tradition of Koen's heuristics into the theories and practices of engineering.

AFTERWORD: TOWARD PROGRESS IN STHV

By holding certain values, philosophers and engineers can advance their respective disciplines independently but limit their capabilities to consummate philosophical engagements of engineering. Since progress in STHV requires a prospering body of scholarship on these engagements, the philosophers and engineers who seek careers in STHV will have to give up some of their values and acquire new ones. In doing so, these philosophers and engineers will trade a portion of their identities with their respective "old" disciplines for a new identity in STHV. At the same time, they must create a substantial knowledge base in which they and their successors have equal shares.

Until now, much of the knowledge base has been the work of philosophers and engineers. Now, there is a need for contributions by philosopher/engineers. Bold new initiatives are required, moreover, to recruit scholars such as these into STHV and keep them there. The financial incentives would likely be enormous. The work assignments might include facilitations of productive discussions among philosophers and engineers; development of core requirements in engineering; but primarily cultivation of scholarship on the philosophical engagement of engineering. A new identity for the STHV scholar will likely evolve out of these bold new initiatives along paths that have already been established.

Chemical engineers were once an assortment of chemists and civil and mechanical engineers. Today, they are a relatively homogeneous group consisting of neither chemists nor civil nor mechanical engineers. A similar evolution can be envisioned for STHV. Thus, a long-term goal might be to create a knowledge base that contains philosophy, engineering, and all their modes of engagement and that can be shared by nonphilosophers and nonengineers.

NOTES

The original version of this paper was read at a science, technology, and human values seminar at Duke University in Durham, North Carolina, on 7 September 1988.

1. R. J. Baum and A. Flores, eds., *Ethical Problems in Engineering* (Troy, N.Y.: Center for the Study of Human Dimensions of Science and Technology, Rensselaer Polytechnic Institute, 1979).

2. The meanings given here follow the outlines of a philosophy of medicine in E. D. Pellegrino, "Philosophy of Medicine: Towards a Definition," *Journal of Medicine and Philosophy* 11 (1986): 10–11.

3. H. Petroski, *To Engineer Is Human* (New York: St. Martin's, 1985).

4. T. H. Broome, "Engineering the Philosophy of Science," *Metaphilosophy* 16 (January 1985): 47–56.

5. B. V. Koen, "Towards a Definition of the Engineering Method," *Engineering Education* 75 (1984): 150–55. See also his *Definition of the Engineering Method* (Washington, D.C.: American Society for Engineering Education, 1985); and his essay in this volume, pp. 33–59.

6. M. F. Rubinstein, *Patterns of Problem Solving* (Englewood Cliffs, N.J.: Prentice-Hall, 1975); see also J. L. Lubkin, *The Teaching of Elementary Problem*

Solving in Engineering and Related Fields (Washington, D.C.: American Society for Engineering Education, 1980, 1981).

7. Broome, "Engineering the Philosophy of Science."

8. Koen, *Definition of the Engineering Method.*

9. P. Caws, "Praxis and Techne," in *The History and Philosophy of Technology,* ed. G. Bugliarello and D. Doner (New York: Harper & Row, 1978), pp. 227–37.

10. S. Florman, *The Existential Pleasures of Engineering* (New York: St. Martin's, 1976).

11. T. H. Broome, "Engineering the Core Curriculum," paper presented at Interface '87: 11th Annual Humanities and Technology Conference, Marietta, Georgia, 22 October 1987.

12. See H. Prochnow and H. Prochnow, Jr., *The Toastmater's Treasure Chest,* 2d ed. (New York: Harper & Row, 1988), p. 168.

13. R. May, *Love and Will* (New York: Norton, 1964), p. 139.

14. T. Kotarbinski, *Praxiology* (Oxford: Pergamon, 1965).

15. See, for example, I. C. Jarvie, "Technology and the Structure of Knowledge," in *Philosophy and Technology*, ed. C. Mitcham and R. Mackey (New York: Free Press, 1972), pp. 54–61.

16. R. Munson, "Why Medicine Cannot Be a Science," *Journal of Medicine and Philosophy* 6 (1983): 183–208.

17. W. Krohn and P. Weingart, "Commentary: Nuclear Power as a Social Experiment—European 'Fall Out' from the Chernobyl Meltdown," *Science, Technology, & Human Values* 12 (Spring 1987): 52–58.

18. M. Martin and R. Schinzinger, *Ethics in Engineering* (New York: McGraw-Hill, 1983).

19. T. H. Broome, "Engineering Responsibility for Hazardous Technologies," *Journal of Professional Issues in Engineering* 113 (April 1987): 139–49.

20. Caws, "Praxis and Techne."

21. Broome, "Engineering the Philosophy of Science."

Deficiencies in Engineering Education

GÜNTER ROPOHL

INTRODUCTION

In 1848, there was a brief bourgeois revolution in Germany that was soon suppressed by resurgent aristocratic regimes. This short period of liberal emancipation was long enough to spawn a variety of democratic movements. For instance, students at the oldest technical university in Germany, the Polytechnic Institute of Karlsruhe, got caught up in the spirit of the times, and a hundred of them left campus and barricaded themselves in village taverns around the city. They declared that they would not return to class until the government of the Grand Duchy of Baden met their demands, one of which was to introduce instruction in economics, history, and philosophy into the engineering curriculum.[1]

Since that time, discussions of the role of the humanities in engineering education have persisted. Time and again demands have been made; and time and again the efforts have run aground because of the narrow-minded orientation of engineering professors, especially in the polytechnic institutes. In each particular instance, it is true, the aims lying behind the demands were different. At times in the past, for instance, it was assumed that engineering had no place in higher education; if it were to gain a place, engineering students would have to be educated in the humanities so that they would be perceived by society to have been adequately educated. More recently, the background assumptions favoring a broader education have changed, and this paper is an attempt to elaborate that point of view.

There are practical and theoretical reasons for a broader perspective. Practically, the challenge comes from ambivalence about the impact of modern technology on society and on the environment—as well as from an expectation that socially responsible engineers will do something about these problems, heading off bad side effects of technological development from the outset. Since the late 1960s in technological settings, two normative approaches have arisen: professional ethics (often couched in terms of the social responsibilities, e.g., of engineers)[2] and technology assessment—usually understood as science-based advice to technopolitical managers.[3] Both approaches, ironically, require preparation on the part of engineers that has not, traditionally, been provided in the engineering curriculum.

Increasing awareness of normative issues in technological practice has been both accompanied and facilitated by a paradigm shift in the understanding of technology. Recent investigations in the history, philosophy, and sociology of technology[4] have made it clear that technology is not so much an area of "autonomous" applied science but a kind of *social practice*. The old—and narrow—understanding of technology regarded technical and societal knowledge as completely different; in this interpretation, the humanities were viewed as largely irrelevant to engineering education. The new paradigm, on the contrary, views as an integral part of technological science societal knowledge—which must therefore be included in the engineering curriculum.

I will follow this outline: in the next section, I will focus on social implications of technological practice; next, I will draw some epistemological conclusions; then I will spell out conclusions relevant to engineering education; and finally, I will argue that technology's normative commitments require that the new paradigm be introduced into the curriculum for future engineers.

SOCIAL IMPLICATIONS OF TECHNOLOGICAL PRACTICE

Usually, engineering is regarded as a mixture of applied sciences and experience-based skills useful in mastering the laws of nature and exploiting natural resources. Although this is not entirely incorrect, it is less than half the truth. Studies in epistemology have demonstrated that there are significant differences between scientific laws and engineering rules; that the theory-application conception is oversimplified in obvious ways.[5] What is more, the whole spectrum of engineering occupations ranges from basic research to marketing, from small-scale design to project management, from civil engineering to computer technology. Suffice it to say that, under such circumstances, the kinds of knowledge required, as well as characteristic work procedures, differ greatly. This contrasts with the conception common to engineering specialties, the illusion that technological practice is no more than the production of artificial objects or products.

In truth, this illusion does not hold up even in the domain most characteristic of engineering in the inventing of solutions for technical problems. If we take a closer look at that activity, we will recognize immediately that what it results in is not just an object but changes in social practice and in the natural environment. With respect to the environmental issue, I can be quite brief because the difficulties are widely recognized. Every technological product requires natural resources in both production and use. When Daimler and Benz invented the automobile, they simultaneously initiated a process of consumption of enormous quantities of steel and petroleum products, not to mention the tolerance of the toxic fumes that escape from engine exhausts—which constitute a large part of air pollution. In the same way, most inventions affect the natural environment, imposing burdens of excessive noise, heat, or wastes. Indeed, by their very existence, technological artifacts have caused lasting changes in the surface appearance of the earth. All of this, perhaps, seemed negligible as long as the numbers were small, but once technological systems came to be used on an ever-increasing scale, it no longer remained possible to classify these impacts as mere side effects; rather, they must be understood to be part of the inventions themselves. In a word, whoever invented the internal combustion engine invented the air pollution that goes with it as well; whoever invented nuclear energy plants invented the problem of storing and disposing of their wastes; and so on. To be honest, engineers long tended to ignore environmental concerns. And they gave no more thought to social and psychological impacts of technology.

Invention does obviously require technical expertise in specific areas. It is equally obviously constrained by experience and by systematized knowledge of physical effects, of material properties, of design principles, and so on. But there is a significant difference between scientific knowledge and technological invention: namely, in the idea of utilization, in the conception of a practical purpose to be achieved by means of the technical device. In the fifteenth century, Gutenberg was almost certainly thinking less about producing movable type than about supplying more people with printed materials or about eliminating the boring job of copying manuscripts by hand. In the nineteenth century, Edison was equally likely to have been less interested in engraving patterns on a tin foil cylinder than he was in recording and reproducing the human voice. In our own day, the fiber optic cable was not invented to demonstrate interesting optical effects but to transmit information at lower cost and with greater efficiency. In principle, then, the process of invention is ruled much more by expediency than by technical feasibility—and expediency is determined by criteria distinct from those of technical procedures or products. In fact, the criteria of expediency derive ultimately from demands of individual and collective practice. Thus every invention means interference in human behavior patterns. Ford not only introduced the mechanized assembly line, but at the same time he contributed to a dramatically changed form of human labor. The inventors of television broadcasting did not only create a technical device for transmitting pictures; they introduced a whole new lifestyle for families. Here again, the social impacts of technology cannot accurately be depicted as mere side effects; they are constituent features of the inventions themselves. Since every invention involves a change in social practice, a concern for social impacts should not be an arbitrary appendage of inventing but an integral part of technological practice.

Most undesired consequences of technology are to be explained by the fact that the majority of inventors and engineers ignore social and environmental concerns. Occasionally they have an implicit sense of recognition of these concerns, but they need to be explicitly aware of the social and ecological constraints on the utilization of technological products—and they need consciously to take into account what is really happening. Invention means intervention—both social-psychological and environmental. Hence introducing new technical products means forming new sociotechnical and ecotechnical systems, whereas engineers have so far generally confined themselves to perfecting technical components, neglecting the sociotechnical and ecotechnical relationships within overall systems. I admit that there have been some beginning attempts to improve the human-machine interface—for instance, in studies of working conditions, in so-called ergonomics, and in industrial design. Such beginnings have not yet penetrated to the core of engineering.

A product or process, once invented, usually needs to be elaborated in advanced design and development stages, and at these stages alternatives for ultimate utilization usually emerge. Therefore, for most engineering problems, there exists more than one solution, and the idea of a one best way in engineering—to be

determined technically—is nothing more than technocratic mystifi-
cation. Every engineer knows that he is constantly making decisions. And when engineers are forced to make an accounting of
their decisions, they readily admit that only some of the criteria
belong to engineering in the narrow sense. There are, of course,
properly technical concerns: feasibility, efficacy, precision, reliability. However, *the* crucial criterion in technical decisions—
whether engineers like it or not—is almost always economic efficiency. Since most engineers work in private industry, technical
practice fits within economic practice. Technical solutions are not
achieved for their own sake but for the sake of business success.
Hence, cost-effectiveness is the prevailing criterion in most engineering decisions.

Although this predominance of economics is part of the engineer's everyday experience, the typical engineer rarely draws the
appropriate conclusions from it. Engineers are reluctant to acknowledge what ought to be obvious—namely, that this economic
constraint on what they do guarantees that, de facto, social conditions and social standards interfere with technical decision making.
As I pointed out earlier, engineering inventions on principle aim at
fulfilling some human purpose; however, that orientation is often
neutralized by the purely formal law of capital accumulation, which
is completely indifferent to considerations of content. As British
businessmen used to say: "Rubbish that sells is not rubbish at all."
Using the more elevated terms of Marx (slightly modified): while
engineers aim at utility values, businessmen are only interested in
exchange values.[7]

There is therefore a fundamental conflict between technical and
economic practice, but engineers, no matter what their differing
experiences, do not acknowledge it; they secretly adopt the perspective of exchange value when they subordinate their work to the
interests of private enterprise. This means—and this time against
Marx—that engineers allow relations of production to govern (their
own) productive forces. Moreover, when engineering decisions are
constrained by economic efficiency, it must be understood that
efficiency depends on costs; costs depend on prices; prices depend
on the relative scarcity of goods; and relative scarcity depends on
social regulations of the distribution of and access to resources. In
consequence, the engineer in choosing the most efficient alternative
is making not only an economic choice but a social choice as well.
This further proves engineering to be an integral part of social
practice.

Inventing, as mentioned, is related to human purposes—and

every purpose is related to some value. Invention implies interference with society and the environment, and the impacts are
either in line with or opposed to specific human values. Furthermore, making decisions about utilizing an invention—even the selection of a particular technical solution—must be made according
to certain criteria, and criteria are also related to values. Consequently, the widespread view that technology is value-neutral is
completely wrong. In today's technological pratice, the technical
and economic values of performance and efficiency predominate.
Nevertheless, as I have shown, the results of technological practice
have an impact on such nontechnical values as environmental
quality, safety, health, welfare, personal development, and social
well-being.[8] So engineering decisions turn willy-nilly into social and
political decisions—with the result that the engineer *must,* in the
end, become aware of the strong political significance of his (or her)
work. (Were the occasion different, I would recall Marx's claim:
"Steam, electricity, and the spinning machine have been more
dangerous revolutionaries than citizens Barbés, Raspail, and Blan-
qui."[9])

CRISIS IN THE ENGINEERING SCIENCES

Contrasting these social and environmental impacts of engineering with the way practitioners of the engineering sciences see
themselves, we are struck with an amazing discrepancy. Even more
amazing is the fact that engineers do not notice the discrepancy—
and so are left completely helpless in facing the real-life demands of
politics and society. Without a doubt, engineers today are totally
incapable of responding to the challenges of professional ethics or
technology assessment. This is what I mean by the crisis in the
engineering sciences—a crisis that, to make matters worse, the
engineers have not even noticed.

We must go back in history to explain this strange turn of events.
It began with the "project of modernity"[10] in the seventeenth and
eighteenth centuries, and the history of the engineering sciences is
no more than a particular instance of a general trend. In a word,
modernity brought about the loss of a worldview. To avoid any
misunderstanding in an age called by some rash intellectuals postmodern, I must emphasize that I recognize the loss of a worldview
as a step forward insofar as such a view was based on the spiritual
and social restrictions of religion and feudalism. Nonetheless, both
society and science had to pay for the new liberty. Increasing

complexity in knowledge and social organization led to what I would call the sectorizing of science and society—their separation into separate sectors. The old worldview was split up, with the separate sectors achieving relative independence—and, to be honest, remarkable effectiveness. In the social realm, sectorizing is manifest in the separation of the private from the public sphere as well as in the appearance of the separate fields of politics, economics, science, and technology (among many others). In science and technology, sectorizing appears as the breaking up of a unified philosophy into specialized disciplines, each of which deals with a single aspect of reality. I cannot give here a summary of Western cultural history, but I would note that the sectorizing process was a matter of historical coincidence and did not follow some supposed law.

This point can be made forcefully by looking at the origins of the engineering sciences. In the late eighteenth century, there were two competing paradigms for a scientific systematization of technical practice—the technological approach properly so called, and the approach by way of applied science.

The first paradigm has been all but forgotten, and I want now to make several comments on it—all the more so since I will be arguing that it ought to be revived. The term *technology* was introduced by the German philosopher Christian Wolff[11] to name "the science of craftsmanship and its products" (which, at that time, obviously meant the science of technical practice). The term was popularized by the German economist Johann Beckmann, who did not know of its origin in Wolff. Beckmann published the first textbook on technology in that sense,[12] and some years later he proposed an outline of a "general technology."[13] It should be granted that Beckmann was, to a certain extent, following the lead of the French Encyclopedists, especially in his lists and descriptions of the crafts and professions. Nonetheless, there was a difference in that Beckmann stressed the interrelationships among craft practices, the economy, and government, following the mercantilist program; also, he went on to generalize about technical phenomena, reducing their diversity to a limited set of unit operations. In this last respect, Beckmann anticipated today's systems approach. To put the matter succinctly, Beckmann was the first to propose a systematic, social science-related paradigm for the engineering sciences. Coincidentally, the paradigm espoused by Beckmann and his followers made a significant impression on Marx, who, in several places in his works, refers to "that very modern science, technology."[14]

This paradigm almost disappeared, the victim of its competitor, the paradigm of technology as applied science. In my opinon, this happened because of the institutional advantages of the latter approach. Moreover, the outstanding historical event was the foundation, in Paris in 1794, of the Ecole Polytechnique. It became the exemplar for all later institutes of technology, reinforcing the separation between technology and the humanities. An example is the first German polytechnic institute at Karlsruhe (mentioned earlier), which self-consciously imitated the Ecole Polytechnique; the founder, J. G. Tulla, had studied in the Paris institute and had refused a chair of engineering education in the humanities-oriented University of Heidelberg.[15] In what followed (everywhere), engineering focused exclusively on applied science and lost, in this theoretical approach, what it could never have lost had it stayed closer to practice—namely, the necessary social-economic dimension of engineering practice. On the other hand, once the separation was complete, the humanities no longer bothered with technology, and the wall of separation between the engineering sciences and the social sciences became absolute. So we should not be surprised that philosophical and social-scientific understanding of technology is badly underdeveloped. Finally, it seems clear that the deficiencies in engineering education addressed here have their roots in deficiencies that affect the engineering sciences.

In order to improve engineering education and make it responsive to the demands of society, we need Beckmann's "general technology"—which in English might better be rendered (because of that language's ambiguities about the term *technology*) as *metatechnology*. To make this improvement, metatechnology ought to be included among the basic engineering sciences. Advocates of the engineering sciences never seem to wonder why this field of knowledge is always described in the plural. Such other disciplines as psychology and sociology have certainly spawned numerous subspecialties, but they have also continued to cultivate a comprehensive view—something like a general sociology or a theoretical psychology. In engineering, by contrast, no such general discipline has been established, even though basing everything on the natural sciences has not been successful. For instance, using physics as a foundation for technological projects has not always enhanced functional performance—without even mentioning that it provides no unifying perspective. Every control engineer knows that certain control functions (e.g., guides for machine tools) can be carried out by mechanical, pneumatic, hydraulic, electrical, or electronic devices almost interchangeably—so that, at least in this field, the traditional distinction between mechanical and electrical engineering has become obsolete. Foundations in areas of physics, though perhaps still important for some optimization purposes, no longer define the actual work an engineer does.

In some limited ways, this has come to be recognized in the establishment of novel engineering fields such as systems engineering and design theory.[16] These approaches do not limit themselves to molar manifestations of physical objects; they start by defining the overall functions to be carried out in proposed technical systems, then look for alternative structures and subsystems, with the design getting to hardware specifications only at the end of the process. In essence, they are following the outlines of Beckmann's "general technology" (or metatechnology); these approaches have outgrown the paradigm of engineering as applied science. Interestingly, opponents of these novel approaches grasp better than their advocates the fact that a paradigm shift has taken place—and they use the standard defenses of silent majorities: institutional suppression, or ignoring the intruders as though they did not exist.

It may be that these hidden enemies believe that these new ideas—of systems engineering and design theory—must be nipped in the bud precisely because they are open to "the normative turn" in thinking about technology; open too, in principle, to the inclusion of social and environmental concerns. Moreover, these approaches reflect explicitly on the nature of engineering methods, taking into account the total life cycle of technological systems, from invention to decommissioning and disposal. These approaches, therefore, necessarily include an attempt to deal with the crisis of the engineering sciences—though, in my opinion, the attempt is still too weak, both in theory and in practice.

We must, consequently, continue to think about metatechnology and its future.[17] Metatechnology must be conceived of in interdisciplinary terms as the study of or the attempt to understand the complexity of the totality of technological practice. Since this is widely recognized as inherently social, metatechnology must implicate, in significant ways, philosophy, the humanities generally, and the social scienes. Moreover, in contrast to specialized disciplinary studies in those fields, metatechnology must be a *generalist* approach, in this sense: it begins with concrete technological phenomena; it brings in and makes use of nontechnical knowledge, but only as related to technological practice; and its use of nontechnical knowledge focuses on specific issues in socio-technical systems. Metatechnology thus takes as its subject matter the structural and functional principles of technological systems both in development and in use.

To illustrate the broad scope of this basic engineering science, I will list briefly some disciplinary perspectives required for the solution of technology problems. Human ecology is needed so that we as a society may study the interrelationships between artificial and natural environments and suggest how technological projects can be appropriately situated in the natural environment. Anthropology is needed to show how technological practice is a natural, evolutionary extension of traditional technical practice, using different means or instruments, and with a special ability to anticipate, evaluate, and act in new ways. Psychology is needed so that we may both understand the sources and conditions of technological creativity and deal with our own attitudes as we face the ever-increasing sophistication of technological systems. Aesthetics is needed so that we may analyze how the senses perceive technological products and discover principles of beauty in technological design. Ethics is needed so that we may establish standards appropriate to technological responsibility, both for producers and users of technological products. Economics is needed so that we may adequately describe technological practice as an effort to increase utility values quantitatively and qualitatively in the face of an advanced stage of the division of labor—with the resultant need to exchange goods, services, and money on an unprecedented scale. Economics is also needed so that we may explain the economic factors in technological development, analyze the often ambivalent relations between efficiency and social welfare, and suggest an optimal level of organization for an economic system under conditions of advanced industrialization. Sociology is needed so that we may explain both the social conditions and the social consequences of technological development—paying special attention to one fact, that social relations in the traditional sense are increasingly being replaced by sociotechnical relations. Policy studies are needed so that we may appreciate the increasing impact of government agencies on technological policies and strategies for development. Political science is needed so that we may study movements demanding popular participation in technopolitical decisions. Legal studies are needed so that we may address problems of the regulation of technological practice by law without putting unnecessary constraints on the dynamics of technical progress. Finally, as the last example, history is needed so that we may study how technology has developed to its present stage, with an eye to detecting regularities in that process.

The cognitive elements I have listed might occasionally be supplied by traditional disciplinary approaches, though they are usually hidden deep within humanistic paradigms far removed from technological culture. Moreover, even if these elements could be teased out of their context and related to engineering, these would be isolated bits of information that make no sense except when integrated within a unified body of knowledge or interpretive system. And here precisely is the challenge of metatechnology (which, it should be noted, is dependent on philosophy of technology). It seems to me that a systems approach—successful already in systems engineering—might provide the appropriate methodology for completing this task of integration. In a word, systems thinking could be the best answer to the sectorizing of knowledge. Scientific specialization cannot be undone, but it—and the entire analytical approach of "the project of modernity"—might find its compensating complement in the synthetic approach, in a systemic holism. In terms of philosophy, an endeavor of that sort would both continue the project of modernity, using better intellectual tools that would allow us to revive a worldview of a more sophisticated sort, and also revitalize our understanding of social practice, which always involves a totality and cannot be limited to a sectoral approach.

IMPLICATIONS FOR ENGINEERING EDUCATION

Having demonstrated that technological practice is at the same time social and environmental practice, and having proposed the outlines of a metatechnology as the science that generalizes from and comprehends all technology-related knowledge, it is time to contrast these ideas with the reality of present-day engineering education. My background experiences, it should be noted, have been formed within a German context. However, from what I know about engineering curricula elsewhere—including the United States—the situation does not seem to be very different.[18]

The largest part of the engineering curriculum is devoted to two broad areas, scientific fundamentals (so-called) and engineering applications. The basic engineering sciences include applied mathematics, mechanics, thermodynamics, electric circuit theory, the physics and chemistry of materials, and so on. Generally these courses are concentrated in the first half of the curriculum, which reinforces the idea that engineering is based on science—and which, incidentally, often kills the motivation of students with a bent for practical problem solving. Only in the second half of the curriculum do students get acquainted with real engineering problems and learn how to do calculations with respect to or to design different types of technical systems. Some particularly sensitive students occasionally notice that there is not much scientific rigor, of the sort they learned in the earlier years, in engineering applications; nonetheless, the ideological identification of engineering with applied science remains as a constant throughout their studies.

In design classes, it is usually the case that only the most advanced, the most successful solutions are presented; this suggests the idea of a "one best way" in engineering, which must be followed rigidly. The large number of earlier trials—and errors—are suppressed, subordinated to the ideology of progress, even of inevitability. Instead of focusing on the ambiguities, coincidences, and failures necessarily attendant on practical problem solving, the manner of instruction gives the impression that designing and technological planning requires no more than the careful following of prescribed recipes. Accepted solutions for problems are presented as unchallengeable, like scientific laws—and all the more so because the context of discovery and invention is totally neglected. (The same is true, of course, in science as well as in engineering education.) Since the process of problem solving is never discussed, the importance that making decisions has in that process is also ignored. All these features of engineering education work together to produce the illusion that technological practice is completely value-free.

A small number of nontechnical courses are permitted, it is true. However, and here the situation is even worse in the United States, these nontechnical course mostly make up for deficiencies in preparation at the high school level; or else they are limited to introductions to various fields under the heading of "general education." This same thing applies even in general studies programs, because such programs typically cover history, law, literature, and the fine arts rather than providing a deeper level of understanding of technology and its human and social consequences. This "additive" approach to general education simply puts a layer of humanities courses on top of engineering courses without any attempt to integrate technical with nontechnical knowledge. As a result, where they are allowed into the engineering curriculum at all, the humanities and social sciences tend to be downgraded, to be thought of as disconnected, as mere appendages rather than an integral part of engineering education. This distortion affects engineering professors as much as students.

In consequence, the whole structure of the engineering curriculum confirms a one-sided view of technological practice; and this view has many characteristics of an ideology in the strict sense— that is, a set of beliefs that does not correspond to reality. On the basis of the foregoing discussion, the elements of this engineering ideology can be summarized as: *technological determinism*[19] (the idea that there is one best way, to be determined technically); *scientism* (the illusion that technology can be reduced to applied science); *internalism* (the idea that engineering is a social institution isolated from others and subject only to its own principles); *positivism* (the claim that technology is value-free); and *historicism* (in Popper's sense,[20] i.e., the idea that technological development is subject to an immanent law of progress rather than to historical contingencies).

Therefore, to improve engineering education requires overcoming this ideology, readying future engineers to take on their social responsibilities. These will, of course, require the gradual introduction of an engineering curriculum that presents a different view of engineering knowledge. Moreover, this paradigm shift in the engineering disciplines must be based on an integrative approach to general education, unifying technical and nontechnical knowledge along the lines of metatechnology. To repeat, systems engineering provides one model for such a program because it demands generalists (but *not* dilettantes) who should be T-shaped persons—broad in their overall knowledge (the horizontal bar of the T), yet deep in one field (the vertical bar).[21] In my view, the horizontal dimension must not be left to chance but must be introduced into the engineering curriculum with all the care devoted to training in a specialty. Incidentally, in order to cope with the problems of fragmentation into ever-narrower specialties, I would maintain that this approach is needed in all fields in higher education. Generalist education may differ widely, but I believe the broad, formal principles are applicable in every field.

One of the most important objections, put forward again and again by advocates of specialist education, can be easily refuted. This is the claim that there is no time available to add new courses to the curriculum; it is crammed full now with technical courses that are essential, and adding more would simply prolong undergraduate engineering education—which is undesirable.

The refutation is this. Nowadays no defender of specialist education can claim that the four-year curriculum supplies all the knowledge acquired over the years in any particular field; it is inevitable that students will graduate with incomplete knowledge (how incomplete is only a matter of degree). Moreover, specialized knowledge quickly becomes obsolete—something that is usually hidden from students. Therefore, it seems to me, it would not be detrimental at all to cut down on the hours required for specialist training, say, by 20 percent, to allow time for general education. Perhaps this percentage seems arbitrary, but to me it seems a reasonable number. Not only should the remaining 80 percent provide enough time for a sound technical education, but 20 percent is a minimum amount of time necessary to provide even basic outlines of the engineering-related general education fields. As it happens, a survey among practicing engineers in Germany turned up exactly this figure: asked what they had missed in their training at the technical universities, they responded that they would have preferred 20 percent nontechnical knowledge.[22] (Some of my colleagues have objected to the precision of this number, saying it represents a superficial approach; but in the mundane business of curriculum planning, there seems to be no other way to go about it.)

In debating these issues, it is of course necessary to spell out the contents of a proposed general education program. In particular, we must be clear about the difference between the old, "additive" approach and a genuinely integrative approach. I think the latter can best be illustrated by examples. For example, instead of requiring a history of economic schools, I would prefer a course on the economics of technological development (one that includes a careful critique of contemporary economic models); and instead of a course on medieval poetry, I would prefer a course focusing on the image of technology presented in modern literature (which would include an analysis of the necessary tension between poetic consciousness and technological reality). Such examples illustrate my point earlier that materials from nontechnical fields must be closely related to technological practice if they are to help provide as comprehensive an understanding of technology as possible. A long list of such course options could easily be derived from the list of metatechnology issues mentioned earlier. Naturally, the reality of in-principle incompleteness, raised as a counter to the argument for specialist training, applies just as much to general education courses. There can be no final list of prescribed or compulsory topics; and in any case the basic goal of general education can be attained by focusing appropriately on just one topic with the right characteristics.

General education, finally, is not a matter of an appropriate content but of an appropriate approach. Its teaching must have it as an explicit goal to acquaint students with the diversity of potential questions and possible answers, to prevent them from coming to believe that issues in metatechnology can be resolved by an appeal to Eternal Truth. Teaching should be seminar-style, with open discussion, and students should be encourged—perhaps even

obliged—to give oral presentations (perhaps giving reports on selected papers). Since most professors also need to learn the generalist attitude, team-teaching should be encourged. When, for
example, a broad-minded engineer and a social scientist argue their
differing views, students get a vivid impression of the legitimate
diversity of opinions. Only in this way can future engineers be
prepared for the cross-disciplinary and political debates that technology involves. Only in this way are they likely to go on and fulfill
the social responsibilities that will be incumbent on them.

CONCLUSION: ENGINEERS MUST FACE THE NORMATIVE TURN IN TECHNOLOGY

In this last section, I return to my opening remarks about recent
normative approaches to technology—namely professional ethics
and technology assessment. Without going into detail,[23] I believe
that neither will succeed in isolation from the other. What I suggest
is a synthesis of individual and collective responsibility—perhaps
under the title "joint technology assessment." The point is that
individuals will not be able to carry out their responsibilities if they
are not supported by relevant institutions, and that technology
assessment institutions will fail unless they are supported by conscientious individual engineers. Though important, engineers are
not the only subjects of technological responsibility; and, in order
to carry out their part of it, engineers must be adequately prepared.

Responsibility in the area of technological practice requires that
engineers be aware of the expected consequences of their activities;
that they know the values to be taken into account; and that
harmful consequences be avoided and beneficial ones fostered. But
how can engineers do this if they do not understand that technological practice involves more than producing artifacts? And how
can engineering students know that if they have not been told? As a
matter of fact, engineering students following the current curriculum learn nothing about the social and psychological or the environmental implications of technological practice; they learn
nothing about social values; and they learn nothing about appropriate procedures for intervening in crucial situations. To make the
matter clear one more time: I make no pretense that the engineer is
a scientific superhuman; the engineer's cognitive resources are as
limited as those of anybody else. For example, I would not expect
an engineer to carry out psychological research on the harm television does to children; but I would expect an engineer working on

the development of small LCD portable TV sets to anticipate the possibility of misuse, to consult with psychologists and educational specialists, and to do whatever is possible to educate the public. To do those things, the engineer has great need of the general education skills dealt with in this paper. It makes no sense to demand ethical professionals or to establish ethics codes if future engineers are not going to be educated so as to develop a sound understanding of the sociotechnical and ecotechnical ramifications of their work.

The same applies to technology assessment. We now have about a generation of technology assessment studies, and they have all been more or less reactive—begun only after some technological development was already underway; begun, that is, after engineers had already invested so much effort in making the decisions they had made that they could legitimately fear "technology arrestment." So I would suggest a second generation of technology assessments that would begin at the very beginning, that would inject a value dimension at every point of development that requires a decision, and that would establish the opportunities and risks of the particular development permanently. An isolated technology assessment institution could not do this; scientists and engineers would have to be ready and willing to share information with the institution, making their contribution to the technopolitical discussion. Once again, this is not feasible unless engineers have acquired a sensitivity about the possible consequences of their work, along with a capacity to take part in multidisciplinary cooperation. Only a radically revised engineering curriculum can provide the needed preparation.

In the end, then, deficiencies in engineering education turn into deficiencies in the social control of technology. So engineering education cannot be a matter for polytechnic institutes and industrial corporations to decide on alone; philosophers, social scientists, even politicians must be involved as well. If it is true that our future is theatened if we do not submit to "the imperative of responsibility,"[24] it is also true that reforming engineering education is one place to start.

NOTES

1. J. Hotz, *Kleine Geschichte der Universität Fridericiana Karlsruhe* (Karlsruhe: Müller, 1975), p. 34. For recent discussions of engineering education, see H. Böhme, ed., *Ingenieure für die Zukunft/Engineers for Tomorrow* (Darmstadt: Technische Hochschule Darmstadt; Munich: Moos, 1980).

2. See, for example, P. Durbin, ed., *Technology and Responsibility*, Philosophy and Technology vol. 3 (Dordrecht: Reidel, 1987); and H. Lenk and G. Ropohl, eds., *Technik und Ethik* (Stuttgart: Reclam, 1987).

3. See, for example, A. Porter, et al., *A Guidebook for Technology Assessment and Impact Analysis* (New York: Elsevier, 1980).

4. G. Ropohl, *Eine Systemtheorie der Technik* (Munich and Vienna: Hanser, 1979). A summary can be found in G. Ropohl, "Some Methodological Aspects of Modelling Sociotechnical Systems," in *Progress in Cybernetics and Systems Research*, ed. R. Trappl, et al. (Washington, D.C.: Hemisphere, 1982), 10:525–36; F. Rapp, "Philosophy of Technology," in *Contemporary Philosophy: A New Survey*, ed. G. Fløistad (The Hague: Nijhoff, 1982), 2:361–412; and W. Bijker, T. Hughes, and T. Pinch, eds., *The Social Construction of Technological Systems* (Cambridge: MIT Press, 1987).

5. F. Rapp, ed., *Contributions to a Philosophy of Technology* (Dordrecht: Reidel, 1974); G. Banse and H. Wendt, eds., *Erkenntnismethoden in den Technikwissenschaften* (East Berlin: VEB Verlag Technik, 1986).

6. The idea was espoused explicitly by F. Taylor in *The Principles of Scientific Management* (1911; reprint, New York: Harper, 1947).

7. Karl Marx, *Das Kapital*, 1.1 (1867).

8. These values were suggested in the first draft of a VDI guideline, *Empfehlungen zur Technikbewertung* (Düsseldorf: Verein Deutscher Ingenieure, 1989); VDI guidelines amount to preliminary forms of national standards. The guideline is reprinted in *Schlüsseltexte zur Technikbewertung*, ed. G. Ropohl, W. Schuchardt, and R. Wolf (Dortmund: Institut für Landes- und Stadtentwicklungsforschung, 1990), pp. 261–86.

9. Karl Marx, "Speech at the Anniversary of the 'People's Paper,' on April 14th, 1856, in London," in Marx, *Ausgewählte Schriften*, vol. 1 (East Berlin: Dietz, 1966), pp. 331–333 (English retranslated).

10. J. Habermas, "Die Moderne—ein unvollendetes Projekt," in *Kleine politische Schriften* (Frankfurt: Suhrkamp, 1981), pp. 444–64; see also Habermas's *Der philosophische Diskurs der Moderne* (Frankfurt: Suhrkamp, 1985).

11. C. Wolff, *Logica* (Frankfurt, 1740), p. 33; cited by J. Guillerme in "Les Liens du sens dans l'histoire de la technologie," in *De la Technique a la technologie* (Paris: Centre Nationale de la Recherche Scientifique, 1984), pp. 23–29; titles translated from the Latin by the author.

12. J. Beckmann, *Anleitung zur Technologie* (Göttingen: Vandenhoeck, 1777).

13. J. Beckmann, *Entwurf der algemeinen Technologie* (Göttingen: Röwer, 1806).

14. Marx, *Kapital*, 1.13. See Dietz ed. (East Berlin: Dietz, 1959), pp. 392 n. 89, 510; my translation.

15. Hotz, *Kleine Geschichte*, pp. 12–13.

16. See, for example, R. Machol, ed., *System Engineering Handbook* (New York: McGraw-Hill, 1965); W. Eder and W. Gosling, *Mechanical System Design* (Oxford: Pergamon, 1965); F. Hansen, *Konstruktionswissenschaft* (Munich and Vienna: Hanser, 1974).

17. For details, see Ropohl, *Eine Systemtheorie der Technik*.

18. I can report on this from personal experience. I was invited by B. Mikolji to participate in a workshop, "Integration of Social Sciences and Humanities into Higher Technical Education," as part of the ASEE conference, Images for the Future, held at the Rochester Institute of Technology in 1983. (My thanks to

Professor Mikolji for the invitation.) Issues very similar to those in Germany were a principal topic of discussion.

19. For details, see G. Ropohl, "A Critique of Technological Determinism," in *Philosophy and Technology,* ed. P. Durbin and F. Rapp (Dordrecht: Reidel, 1983), pp. 83–96; also, Ropohl, "Friedrich Dessauers Verteidigung der Technik," *Zeitschrift fur Philosophische Forschung* 42 (1988): 301–10.

20. K. Popper, *The Poverty of Historicism* (London: Routledge and Kegan Paul, 1960).

21. R. Machol, "Methodology of System Engineering," in *System Engineering: Handbook,* ed. R. Machol, pp. 1-3–1-13.

22. H. Hillmer, R. Peters, and M. Polke, *Studium, Beruf und Qualifikation der Ingenieure* (Düsseldorf: VDI-Verlag, 1976).

23. For details, see G. Ropohl, "Neue Wege, die Technik zu verantworten," in *Technik und Ethik,* ed. Lenk and Ropohl, pp. 149–79. An English version will be published with the title, "Individual and Institutional Responsibility in Technical Practice," in *Normative Issues of Technical Development,* ed. I. Hronszky (Budapest: 1988), pp. 41–59.

24. H. Jonas, *The Imperative of Responsibility: In Search of an Ethics for the Technological Age* (German original, 1979; Chicago: University of Chicago Press, 1984).

Index